Velocity (m/s)

1 foot/sec (ft ... /s
1 mile/hour (... m/s

Volume (m^3)

1 barrel (= 42 U.S. gallons)	$= 0.15899$ m^3
1 fluid ounce (U.S.)	$= 2.9574 \times 10^{-5}$ m^3
1 gallon (U.S.)	$= 3.785 \times 10^{-3}$ m^3
1 liter	$= 10^{-3}$ m^3
1 teaspoon	$= 4.9289 \times 10^{-6}$ m^3

Some Important Constants

acceleration due to gravity (standard)	g	9.81 m/s^2
Avogadro constant	N_A	6.022×10^{26} kmol^{-1}
Boltzmann constant	k_B	1.381×10^{-23} J/K
Planck constant	h	6.626×10^{-34} J s
speed of light in vacuum	c_0	2.9979×10^8 m/s
Stefan–Boltzmann constant	σ	5.67×10^{-8} W/(m^2 K^4)
universal gas constant	R_u	8.314×10^3 J/(kmol K)

Properties of Air and Water at 15°C (288 K) and One Atmosphere Pressure

		Air	Water
molecular weight (kg/kmol)	M	28.97	18.02
density (kg/m^3)	ρ	1.22	1.00×10^3
kinematic viscosity (m^2/s)	ν	1.45×10^{-5}	1.14×10^{-6}
thermal conductivity (W/(m K))	k	2.51×10^{-2}	0.59
thermal diffusivity (m^2/s)	α	2.05×10^{-5}	1.41×10^{-7}
specific heat at constant volume (J/(kg K))	c_v	0.718×10^3	4.173×10^3
specific heat at constant pressure (J/(kg K))	c_p	1.005×10^3	4.186×10^3
enthalpy of vaporization (J/kg)			2.465×10^6
coefficient of volume expansion (K^{-1})	β	3.48×10^{-3}	1.5×10^{-4}

Properties of the atmosphere are given in Table 1, Chapter 5

Aimed at beginning engineering students, this book presents the basic ideas of thermodynamics, fluid mechanics, heat transfer, and combustion through a real-world engineering situation. The engine is related to the atmosphere in which it moves and exhausts its waste products. In addition to the traditional thermal–fluid topics, the book includes a chapter on the atmosphere, with a particular focus on the greenhouse effect and atmospheric inversions. The social implications of engineering in a crowded world with increasing energy demands are also addressed. This novel approach will capture the reader's attention and set the stage for subjects that will be studied later in greater depth.

Students in mechanical, civil, agricultural, environmental, aerospace, and chemical engineering, as well as the physical sciences, will welcome this engaging, well-illustrated introduction to thermal–fluid engineering.

AN INTRODUCTION TO THERMAL–FLUID ENGINEERING

AN INTRODUCTION TO THERMAL–FLUID ENGINEERING

THE ENGINE AND THE ATMOSPHERE

Z. WARHAFT

Cornell University

PUBLISHED BY THE PRESS SYNDICATE OF THE UNIVERSITY OF CAMBRIDGE
The Pitt Building, Trumpington Street, Cambridge/CB2 1RP, United Kingdom

CAMBRIDGE UNIVERSITY PRESS
The Edinburgh Building, Cambridge CB2 2RU, United Kingdom
40 West 20th Street, New York, NY 10011-4211, USA
10 Stamford Road, Oakleigh, Melbourne 3166, Australia

First published 1997

Printed in the United States of America

Library of Congress Cataloging-in-Publication Data
Warhaft. Z. (Zellman). 1944–
An introduction to thermal–fluid engineering: the engine and the
atmosphere / Z. Warhaft
p. cm.
ISBN 0-521-58100-1 (hc). – ISBN 0-521-58927-4 (pbk.)
1. Heat engineering. 2. Fluid dynamics. 3. Internal combustion
engines – Exhaust emissions – Environmental aspects. I. Title.
TJ260.W217 1997 97-89
621.402 – dc21 CIP

*A catalogue record for this book is available from
the British Library*

ISBN 0 521 58100 1 hardback
ISBN 0 521 58927 4 paperback

For Zoe and Simon and to the memory of Sacha

Contents

Preface

Frequently, engineering students confront me with the question: "All of this math, physics, thermodynamics, and fluid mechanics is fine, but when are we going to study real engineering?" And then sometimes, after a pause: "Anyway, what *is* engineering?" My answer to these students, who are often well into their program, is usually an evasive mumble: "Wait a little longer. It will all come together. You will see." The students walk away none the wiser, and as the years go by I find my reply more and more dissatisfying. After all, law and medical students know what their subject is about, and so do upper-level physics and chemistry majors. What is wrong with the way we educate our engineers?

Consider for the moment the way we teach physics. By the end of their freshman year most students know what the landscape of physics looks like. Courses are connected in sequence, each building on the other: mechanics, electrodynamics, optics, and so on up to quantum and relativity theory. By the time physics students have reached their second or third year, they have a feeling for the progression of the subject and how the pieces fit together. They have also acquired technique in problem solving. Although they have not yet plumbed the depths, the first introductory overview provides them with substantial intellectual apparatus and hence a sound basis for a deeper second look later on. This progression is primarily due to a strong tradition of texts and elementary courses that cover the whole field at an introductory but nontrivial level. Engineering, like physics, is a discipline with a logical and coherent progression, yet my bookshelf is bereft of any introductory, unifying text. This lack, I suspect, is a major reason for the perplexity of many engineering students and provides much of the motivation for writing this book.

Engineering is difficult. Unlike the natural sciences, in which the major objective is to analyze the physical or biological world, engineering encompasses both analysis and synthesis (or design, as it is more often called). Moreover, engineering is a social activity: it does not progress as a purely intellectual or puzzle-solving pursuit. Airplanes, automobiles, and rockets change our environment and our social behavior; they make our wars more potent and devastating and affect our economy. No serious engineer can be oblivious to these issues. In writing this book, I have kept the social implications foremost in my mind. Apart from their intrinsic importance, they also provide the focus needed to unify the various topics.

My objective, then, is to provide an introduction to engineering by laying out the foundations in a coherent and connected way. Because engineering is a broad discipline, I cannot be completely general. I have confined myself to the thermal–fluids area of engineering, which constitutes about half of the mechanical engineering syllabus and a considerable portion of chemical, civil, aerospace, agricultural, and environmental engineering. So although this book is not as comprehensive as an introductory physics text, it is nevertheless broad enough to provide the overview that I believe is lacking.

My development of thermodynamics, fluid mechanics, heat transfer, and combustion is aimed at giving beginning students, at the first- and second-year college level, a firm foundation so they can do calculations and solve problems. When these topics are seen again later on, as separate subjects, students will be at home with their basic concepts and will not be bewildered by the details or how the subjects interrelate. I hope that this book may also be used by engineering nonmajors who wish to have a detailed overview, much in the spirit that physics nonmajors study introductory physics. The prerequisites for this book are concurrent first-year calculus and physics courses.

Teachers of the thermal and fluid sciences will see that some of my topics and the way they are developed are relatively standard, whereas others are not. Thus, the chapter on thermodynamics follows the familiar path first provided by Keenan, whereas the chapters on fluid mechanics and heat transfer depart considerably from the usual method of introduction. Here, at the outset, I accentuate the role of viscosity and flow losses. Turbulence, which is responsible for the major flow losses as well as for enhancing mixing and combustion, is discussed in some detail. Although not mathematical, the method of thinking is different and requires considerable patience. But so too does classical thermodynamics, a subject that has been dished up, without apology, to beginning students for decades. By discussing turbulence, I introduce the students to approximate methods of analysis as well as to the role of computers in engineering. Both these aspects play an ever increasing and important role in modern engineering. What I believe to be wholly new in the book, however, is the way I connect the subjects by relating the engine – its thermodynamics, fluid mechanics, heat transfer, and combustion – to the environment in which it moves and exhausts its waste products. To make this connection, I have included a chapter on the atmosphere, with particular focus on the greenhouse effect and atmospheric inversions. Here the student will also see that atmospheric science is really an application of engineering fluid mechanics, heat transfer, and so on to a particular system. So apart from showing that the engine is not an isolated object but must be studied in relation to the environment, I also show that the tools used to analyze the engine are indeed the same as we use to understand the atmosphere and its motion.

In summary, this book is a holistic introduction to the thermal–fluid side of engineering. Its aim is to lay out the foundations in a coherent manner, providing building blocks for further courses. Although written primarily for freshmen and sophomores, I hope it will also serve as a supplement to the upper-level courses. The book has the broader aim of showing that engineering is not merely technology but encompasses social and moral questions, making it perhaps the most fascinating profession of our age and one with the profoundest influence.

Acknowledgments

My gratitude goes to the following who read parts of the manuscript and provided me with generous and thoughtful comments: Dr. David Curzon, Dr. Thomas Dreeben, Prof. Elizabeth Fisher, Prof. Phillip Holmes, Frederick Horan, Prof. John Lumley, Prof. Frank Moon, Laurent Mydlarski, Dr. Peter Taylor, Prof. William Streett, Prof. Kenneth Torrance, and Dr. Marcela Villarea. The drawings were done with great care by Ali Avcisoy, and the numerous drafts were cheerfully typed by June Meyermann and Debbie DeCamillo. Prof. David Caughey (director of the Sibley School of Engineering) and Michelle Fish (director of A.P. Sloan Foundation grant for Women's Programs in Engineering at Cornell) provided me with some financial assistance. I am most thankful to all these people for their help.

Most important, I thank Gail for sharing my convictions and for gently attempting to dock my prolixity; and our children, who provide the reason for it all and to whom the book is dedicated.

1

Introduction

1.1 Why Is Engineering Different?

We all have a great desire to think big, be it in terms of career plans, travel adventures, fame, or financial gain. TV programs aim to reach the largest audience, and religion attempts to be universal. It would almost seem redundant to include science here, because its very foundation is one of universality: would we even want to consider a physical theory in which one set of laws held in New York and another in New Zealand? Universality is the very essence of physical theory; not only must it hold in every location, but it should cover as large a scope or be as general as possible. This is why Einstein's theory of gravity is an improvement on Newton's. The latter fails for the strong gravitational fields that occur close to stars and other celestial objects.

Engineering too, we should surely say, aims at universality. After all, a bridge in the Eastern United States obeys the same laws of statics and elasticity as a bridge crossing a remote gorge in Central Asia. True, but the object will turn out to be very different in the two places because of climate, terrain, availability of materials, and type of traffic. One may be required to be a six-lane concrete and steel freeway, the other, a rope suspension bridge (Figure 1.1). Both may be equally vital to the respective communities. The same comments may be applied to building design, power generation, and even communication equipment; the physical laws are the same but the engineering realization should be different from region to region and country to country. We see immediately a strong distinction between physics and engineering. Physical theory, in its best form, is devoid of context, whereas engineering, in its best form, is firmly placed in location, taking into account climate, local materials and terrain, economy, and social structure. When I say "best form" I think of the opposites: physical theory that is location-dependent or a bridge designer who sells the same design everywhere. This example also shows that engineering is, by its very nature,

Figure 1.1 A modern concrete highway and a rope bridge. Although the broad physical principles of elasticity are the same for each bridge, the engineering realization depends on the location, type of use, availability of materials, and so on. Engineering is firmly rooted in society and the environment. (Upper photo reprinted with permission of HNTB Corporation. Lower photo reprinted with the permission of Ewing Galloway, Inc.)

environmental and social. It is strange, given the strong distinction between engineering and the physical sciences, that the two disciplines attract similar types of students. It is also strange that one is more likely to find the physics department in the center of the campus, near the arts college, where the students and faculty can mingle with sociologists, psychologists and the like, whereas engineering schools tend to be isolated, on the edge of the campus. Not always, of course, but often. We will return to these paradoxes in Chapter 7.

Apart from the strong environmental and social aspects, engineering is distinct from physics and the other sciences from which it derives its basic laws in another unique characteristic, that of design. Physicists develop theories about the world. Traditionally this has been the natural world of atoms, planets, and stars, but it can also encompass constructions such as transistors or liquid crystals. The main point is to provide a coherent explanation of how these objects behave the way they do. Engineering, on the other hand, is not always so finicky about fundamental understanding. For example, the processes inside an internal combustion engine are still not completely understood, both in terms of the combustion processes themselves and the related fluid dynamics and thermodynamics. This, however, does not stop engineers from developing more efficient and less polluting ones. These newer engines are usually more complex, and their design would be easier if the fundamental principles were better understood, but engineers cannot wait; there are social, economic, and environmental pressures to move ahead. On the other hand, the tendency in physics is to understand the simple before going on to the complex (Figure 1.2). Engineering design, it turns out, is almost invariably done without a complete understanding of the physics (or chemistry) of the intended object, be it an air filtration unit, a car, or a supersonic transport. This lack of knowledge provides excitement as well as frustration. Notice that the design aspect of engineering is intimately linked with the social (and environmental) aspects because all design is, in some way, the result of social needs.

1.2 Thinking Big and Small

If everyone in your class, or indeed college, were to give you a penny, your fortunes would not be noticeably changed. However, if everyone in the United States were to do so, there would be no need for you to go on studying, at least for the sake of providing a future livelihood. Your fortune would have been made. Very small numbers multiplied by moderate numbers produce small numbers. Very small numbers multiplied by extremely large numbers produce very large ones.

The multiplication of very small and extremely large numbers is a key aspect in both the production and the environmental aspects of modern engineering. Consider first production. In earlier times, when, for example, a car plant may have produced a few hundred cars per year, the loss of a few ounces of brass or

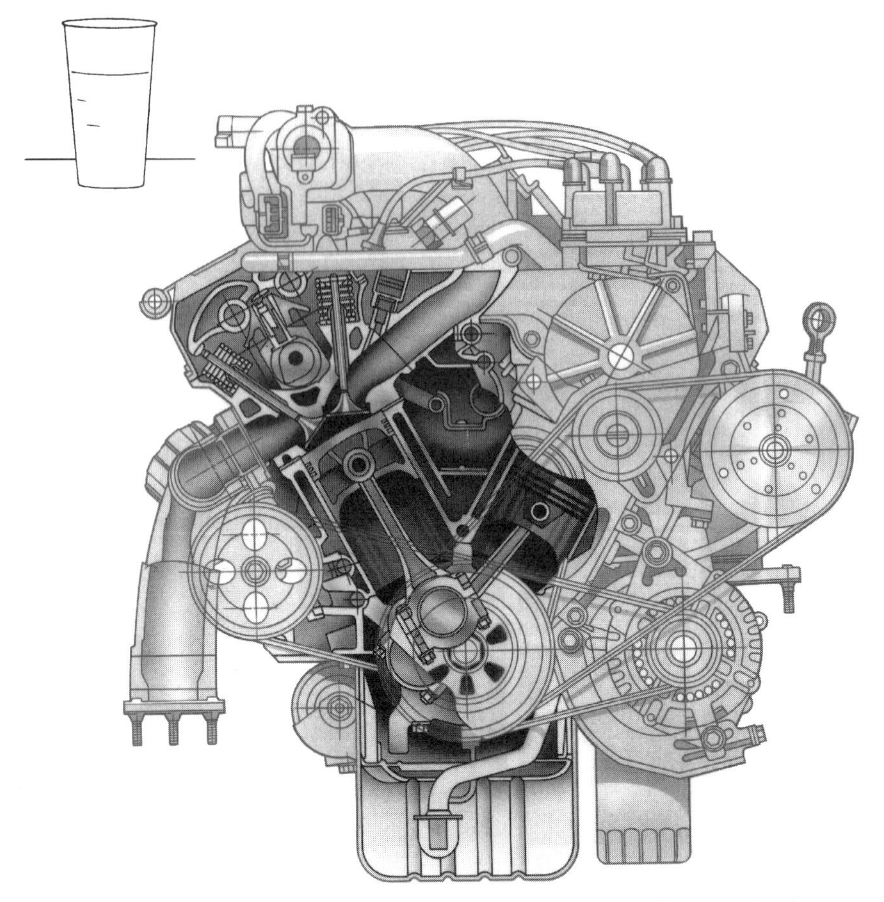

Figure 1.2 A modern automobile engine and a glass of water. Engineers design complex systems, sometimes without fully understanding the details of how they work. Scientists analyze simpler phenomena in great detail, for example, the structure of liquids (such as water) and their interaction with solid and gas interfaces. (Reprinted from a 1994 advertising brochure with the permission of the Lincoln-Mercury division of Ford Motor Co.)

some other relatively costly metal in the construction of a particular component could go unnoticed, whereas today, with a plant producing five million cars per year, this loss would amount to hundreds of tons of material. You might say that the amount per car is surely what matters, but this is not the case, because as numbers become larger we begin to feel a new and vitally important constraint: the finite size of the world and its resources. This constraint forces us to consider absolute rather than relative values. The amount of metal the small car-producer used may have been unnoticeable in terms of world resources; the large car-producer may be using a small but significant amount of the total world production and will be forced to consider material losses in a new light.

The same reasoning applies to pollution; again we use the car for our example. As a child in the fifties, when there were still comparatively few cars, I often saw the most deplorable behavior of dumping engine oil down storm drains or at the end of the properties, on waste land. (In those days many people used to change their car oil themselves.) I grew up in Australia, a large underpopulated country, and the effect on the environment was small. Today there are over 500 million road vehicles in the world; if only 10 percent of them were to dump their used oil, this would amount to about fifty million gallons. The massive oil spill in 1989 off the coast of Alaska from the tanker The Exxon Valdez was ten million gallons. Of course the tanker spill was confined to a single location, so the destructive effects were greatly amplified, but so too, to a limited extent, is oil dumping, because most of the world's cars are in Europe and on the coastal regions of North America.

The problem of very small numbers multiplied by extremely large ones is unique to modern engineering because it is only now that the very large numbers resulting are of worldly size. We will return to this concept repeatedly.

1.3 Crude and Refined Calculations

Consider these questions. How much energy of motion is converted into heat because of internal friction in a cumulus cloud? What is the per capita consumption of energy in the United States compared to Guatemala? What is the overall maximum efficiency that can be obtained from an automobile? Consider now a second set of questions. What is the drag coefficient of a Boeing 757 at cruise altitude? What is the maximum torque of the drive shaft of a particular type of power generator running at full load? What is the gravitational constant?

Both sets of questions require quantitative answers, but in the first case only ballpark figures are suggested, whereas for the second set of questions precise answers are needed (Figure 1.3). Thus, all cumulus clouds are different, but it is of interest in atmospheric energetics to know whether the typical power, due to its motion, is watts, kilowatts, or tens or hundreds of kilowatts. This type of information will be used to determine the overall energy budget of the atmosphere and compared, say, with the latent heat released because of condensation in the cloud. Here answers to within an order of magnitude (i.e., is it 10 or 100 kW?) may be satisfactory. Similarly, knowledge of the relative energy consumption of a first-world country such as the United States compared to a small, war-torn central American country, such as Guatemala, is needed for predicting future energy needs and trends, but here too the type of answer required is approximate: is it ten to one or twenty to one? Answers to within 20 or 30 percent will usually be good enough. After all, our ability to predict future trends is not accurate, and so it would be a waste of effort attempting to answer this question precisely.

Figure 1.3 A jet airliner and a cumulus cloud. Calculations of the power dissipated by the motion of the cloud are needed in order to determine the overall energetics of the atmosphere; the drag of the aircraft is needed in order to determine its fuel consumption. The power dissipated can be determined only crudely, whereas the drag must be determined to high accuracy. Both types of calculations are important. In this book we will be mainly concerned with approximate calculations. (cloud photo: NCAR. Airplane, courtesy of Boeing Airline Magazine.)

Finally, and again for energy projections: Is the typical overall efficiency of a car 70 percent, 30 percent, or 15 percent? An answer to within 1 or 2 percent to such a general question is meaningless because loads, usage, and other factors vary greatly. Notice that ballpark answers provide a great deal of information, because they give a quantitative feel to the problem, and from this one can decide whether the quantity is important, whether it can be changed or improved upon, or whether differences are significant. These crude questions are often done on the back of an envelope, but they can require subtle and original thinking. Engineering students tend to shy away from them because their solutions lack formal techniques.

The second set of questions is more familiar to engineering students; most of the problems in engineering texts are of this type. They require a formal methodology and very precise answers. Thus the drag coefficient of a particular aircraft must be known to within a few percent, and as all airlines know, a reduction of 1 or 2 percent can result in millions of dollars in savings in fuel for a full fleet. Similarly, the maximum torque of a drive shaft must be precisely known; its safety and reliability are contingent on this number. Finally, for satellite and rocket orbital calculations, knowledge of the gravitational constant to within a factor of two would be disastrous. Presently it is known to within an error of about 0.1 percent, and even this level of accuracy is not considered great enough.

This book will be concerned mostly with what I will call crude or ballpark types of calculations rather than with refined or precise calculations. My reason is twofold. First, crude calculations often provide great insight into new questions. There is nothing humdrum or routine about them, and the answers are often exciting in themselves. For instance, simple dimensional reasoning shows that a large cumulus cloud could dissipate tens of megawatts of energy. Immediately we begin to compare the cloud with atom bombs and power stations. The actual number provokes divergent and creative thinking. Second, modern engineers are beginning to play a greater role in multidisciplinary activities. These include policy analysis in areas such as the environment and arms control, and the planning of alternative energy and transportation systems. In all of these activities the engineer will have to bring together knowledge from a number of different disciplines. Calculations will by necessity be approximate because of the uncertainty of the relative weighting of the various parameters. They will also be difficult, not because they will be highly complex or mathematical, but because they will require divergent thinking of the broadest kind. Such is one of the challenges of the modern engineer.

1.4 The Engine and the Atmosphere – A Broad Outline

So far, I have indicated the flavor of my approach to engineering. Engineers, by the very nature of their profession, must be involved with society and the environment. They should be good at doing calculations, particularly the approximate type, which can reveal a great deal with a relatively small investment. They should be interested in the significance of their answers, and they should be adept at relating these answers to the problem as a whole. I will now outline in a broad way the subject of the book. It is my desire that in the subsequent chapters the reader not get lost in the details; I am trying to present a coherent account, a story, and here I will give an outline, stripped of the details. From time to time the reader may wish to refer back to this section in order to gain perspective.

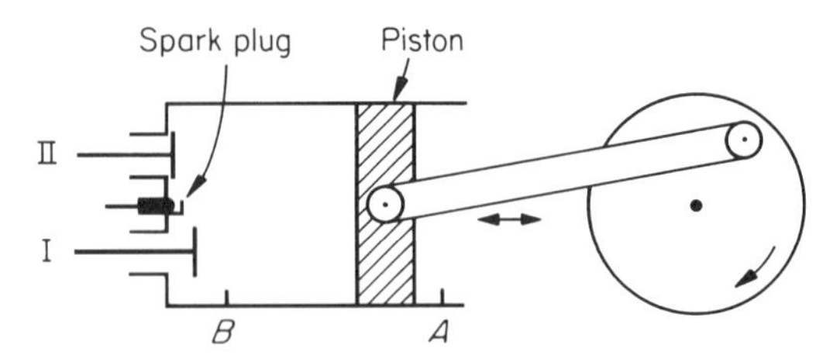

Figure 1.4 A single-piston spark-ignition internal combustion engine. I is the intake valve (for fuel and air), and II is the exhaust valve. The sequence of events that produce the work is explained in the text. Notice the piston and cylinder in the complex engine of Figure 1.2.

1.4.1 The Engine

There are approximately 500 million road vehicles in the world, each with an engine – a machine that converts energy (usually chemical energy from hydrocarbon fuels) into useful work. Terms such as **engine**, **energy**, and **work** will be more rigorously described in subsequent chapters; here I will rely on a combination of your intuitive notions as well as your acquaintance with these concepts from physics and chemistry. In road vehicles, the engines are mostly internal combustion (IC) piston-cylinder engines, whereas in power plants and jet aircraft they consist of compressors and turbines. For the moment we will concern ourselves with the piston-cylinder engine. A drawing of a real single-cylinder piston engine is shown in Figure 1.4. I will describe what is happening as it goes through its routine.

The piston (Figure 1.4) moves back and forth, or undergoes a cycle, with monotonous repetition and is coupled to mechanical devices (on the right of the drawing) that can change back and forth motion (or up and down motion, depending on the orientation) into rotational motion, thereby turning a wheel which can then push a car or lift a weight. I will not be particularly concerned with the mechanical coupling; my main concern is that the back and forth motion in the cylinder can in the end do work, which we will define as the lifting of a weight. Our main concern will thus be with the piston-cylinder arrangement itself.

Begin the cycle with the piston far to the left (position B, Figure 1.4) with the bottom valve, I, open. The piston is allowed to move to the right, drawing in a mixture of air and fuel through the valve. When the piston has reached the far right (position A), valve I is closed and the air–fuel mixture is compressed by the piston, which moves back toward the left. When the piston has moved almost back to B, a timing device allows the spark to occur, igniting the air–fuel mixture. The pressure in the cylinder increases dramatically, and the piston is forced to the right doing work on the surroundings. This is known as the expansion stroke. As the piston approaches A, the exhaust valve (II) is opened and the burned air–fuel mixture is pushed out of the cylinder (the exhaust stroke) as the piston returns to position B. Then valve II closes and valve I opens and the cycle is repeated.

This is roughly how a four-stroke IC engine works. Over the duration of a cycle, the engine does work external to itself (on its surroundings), even though some work is done on the piston cylinder system itself during the compression part of the cycle. (We will show later that the work done during the expansion stroke is far greater than that lost on the compression.) Notice that it is impossible to build an IC engine (that goes through a cycle and does work) without producing exhaust products, because the pressure would quickly build up and soon the engine would blow up if the gases were not exhausted.

Large temperature differences occur in the engine as it goes through its cycle. The subject of thermodynamics (the study of heat and force) was developed in the early part of the nineteenth century in order to understand such engines, in particular to determine the maximum amount of work that can be obtained from a fixed amount of fuel, as serious a consideration then as it is now. In those earlier times, the piston-cylinder system was similar but the way it was energized was somewhat different: It was an external combustion engine, and this is shown in Figure 1.5. Here instead of producing high temperatures inside the cylinder by means of internal combustion, the high temperatures were produced external to the engine and heat was transferred through the cylinder wall to the gas inside. The pressure then rose, and the piston went through its cycle, just like the IC engine. Notice that for the piston to return to the left, the source of heat must be removed and a cool sink (usually the environment) must allow heat to be transferred from the cylinder to the surroundings. Unless this were done the pressure would keep on increasing, and the engine would blow up. In the diagram we have used a hot plate (the heat source) and a cold plate to represent the external combustion and the cool environment, respectively. Often in this book I will have this more idealized type of engine in mind, particularly when discussing thermodynamics (Chapter 2) and the mechanisms of heat transfer (Chapter 4).

The great thermodynamicists of the early nineteenth century, Carnot, Clausius, Joule, Mayer, Lord Kelvin, and others, were not particularly concerned about the pollution from the coal they burned to produce heat, but they were interested

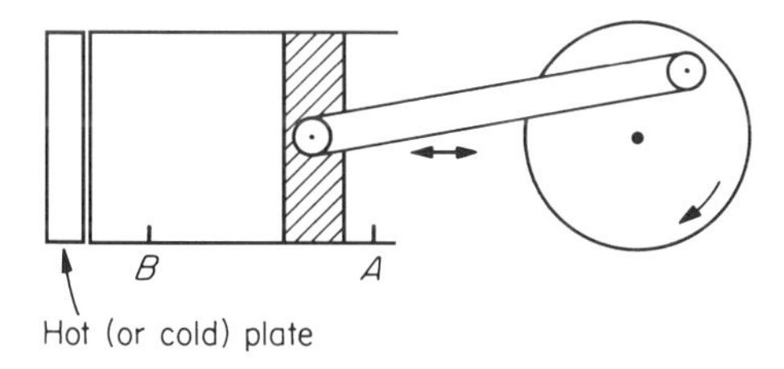

Figure 1.5 A heat engine, with the spark ignition replaced by a hot plate that causes the expansion and the exhaust valve replaced with a cold plate that allows for the gas in the cylinder to return to its original temperature and pressure after the completion of the cycle. Note that the gas in the cylinder is not replaced each cycle, as it is in the IC engine (Figure 1.4).

in having a high-efficiency engine, because coal mining was a difficult and hazardous business. They asked, what is the maximum work that can be achieved from a heat engine (as shown in Figure 1.5) for a given amount of heat input? Can all the heat derived from burning the coal be converted into work? Defining the thermal efficiency, η_{th}, as

$$\eta_{th} \equiv \frac{\text{net work from the heat engine}}{\text{heat into the engine}}, \tag{1.1}$$

their question was, can η_{th} be equal to one? Their answer was no. A little reflection suggests that this must be so. In order for the engine to complete a cycle, some heat must be rejected, that is, not all the heat taken in can be converted to work (both heat and work, we know, being forms of energy; see Chapter 2). What is remarkable is that these early thinkers were able to provide a theoretical upper bound on the efficiency of a heat engine. These issues lie at the heart of thermodynamics and will be dealt with in Chapter 2.

Let us return to the source of heat itself. To produce the high temperatures, it is feasible that the sun's rays could be focused at the bottom of the cylinder, but, as we know, the most likely source of energy will be the burning of a hydrocarbon fuel. Such fuels have molecules consisting of an arrangement principally of hydrogen and carbon atoms (and sometimes oxygen). When they undergo combustion in air, the rearranging of the chemical bonds releases large amounts of energy. For example, the equation for the combustion of octane in air is

$$C_8H_{18} + 12.5(O_2 + 3.76N_2) \rightarrow 8CO_2 + 9H_2O + 47N_2 + 48 \times 10^6 \text{ J/kg fuel}. \tag{1.2}$$

We have included 3.76 molecules of N_2 for every 1 of O_2 because this is the makeup of air. Notice, however, that nothing is happening to the N_2: Here it is only the oxygen that is reacting. This equation has two important aspects. First, carbon dioxide and water are always products of hydrocarbon combustion, be it octane, methane (CH_4), wood (CH_2O), or sugar ($C_6H_{12}O_6$). There may be other products such as CO and nitrogen oxides, but if the combustion takes place in any reasonable way, CO_2 and H_2O must be produced. Second, a lot of energy is released – the chemical energy stored in one teaspoon of octane is approximately the same as the kinetic energy of a brick traveling at 1 km/sec (2,200 mph). This is why fossil fuels are so attractive as an energy source.

Let us assume that C_8H_{18} is representative of gasoline. (In fact, gasoline consists of many different kinds of hydrocarbon molecules, but its average molecular weight is close to that of C_8H_{18}.) Equation (1.2) shows that one molecule of C_8H_{18} produces eight molecules of CO_2. Because the molecular weight of C_8H_{18} is 114 and that of CO_2 is 44, 1 kg of our gasoline produces $(8 \times 44)/114 = 3.1$ kg of CO_2. The density of gasoline is about 70 percent that of water (notice that it floats on water), and because the density of water is 1,000 kg/m^3, 1 liter

(10^{-3} m^3) of gasoline produces $3.1 \times 0.7 = 2.17$ kg of CO_2, or in the U.S. system, 1 gallon (3.785 liter) of gasoline produces 18.1 lb of carbon dioxide. On a holiday trip of 200 miles (320 km) an average car, consuming around 1 gallon in 25 miles, puts approximately 290 lb (132 kg) of CO_2 into the atmosphere for the round trip. This is about the mass of two college students.

Now let us look at this in global terms. Assume that on average, each of the 500 million vehicles in the world burns 2 gallons (7.6 liters) of gasoline per day (trucks and buses burn much more; occasional users, much less). This amounts to 3.65×10^{11} gallons per year. From this, 2.95×10^9 tons (or 3×10^{12} kg) of CO_2 will be produced per year by the combustion process. Note that we could be in error by a factor of two or three, but it is unlikely that our error is much more than this. As discussed in Section 1.2, we are mainly interested in approximate answers; whether it is two or six billion tons of CO_2 is of little significance at this stage; what is of interest is that it is 10^9 and not, say, 10^7 or 10^{12}.

What does this large number mean, and does it matter? In order to answer that question we have to learn about the atmosphere itself.

1.4.2 The Atmosphere

Engines are small; their cylinders have dimensions of the order of tens of centimeters. The atmosphere is very large; its dimensions are measured in thousands of kilometers. Can the exhaust from these engines affect the whole atmosphere? Let us first determine some of its salient characteristics.

What is the mass of the atmosphere? What is its total energy of motion? Answers to these questions are provided in handbooks of geophysics or in the appendixes of meteorology books. But in the spirit of learning how to do quick, exploratory engineering calculations, we will answer these questions using a few well-known facts. Consider the mass of the atmosphere. We know from weather forecasts (and home barometers) that the pressure at ground level, p_0, is around 15 lbf/in^2 or 10^5 Pa (N/m^2). This is the force per unit area exerted on the land, on our bodies, etc., due to the column of atmosphere above us. The total force, F, acting on the earth is thus $p_0 A$, where A is the total area of the earth's surface. This must be equal to $m_a g$, where m_a is the mass of the atmosphere and g is the acceleration due to gravity. Therefore, the mass of the atmosphere is

$$m_a = p_0 A/g = p_0 4\pi r_e^2/g, \qquad (1.3)$$

where $A = 4\pi r_e^2$ and r_e is the radius of the earth. So if we know the ground level air pressure, p_0, the radius of the earth, and the acceleration due to gravity, we can determine the mass of the atmosphere. Both r_e and g can be easily determined; their values are 6.370×10^6 m and 9.8 m/s^2, respectively. Substituting these, as well as the average ground level pressure, into Equation (1.3), we find that $m_a = 5.2 \times 10^{18}$ kg.

At the end of the previous section we determined that the annual production of CO_2 from motor vehicles alone was approximately 3×10^9 tons or 3×10^{12} kg per year. We can now determine the ratio of CO_2 injected to the atmosphere per year by motor vehicles to the total mass of the atmosphere. It is $(3 \times 10^{12})/(5.2 \times 10^{18}) = 5.8 \times 10^{-7}$ or approximately 0.6 parts of CO_2 per million parts of atmospheric mass per year. Does this small amount of CO_2 matter? After all, it is less than one millionth of the mass of atmosphere per year. The answer to this question is very important but very complicated and will be one of the main themes of this book. Before we begin to address it, we must discuss the nature of the atmosphere itself a little more.

The major constituents of the atmosphere are nitrogen, N_2, which has a molecular weight of 28, and oxygen, O_2, which has a molecular weight of 32. The percentage composition by volume is 78.08 percent of N_2 and 20.95 percent of O_2, so together these make up over 99 percent of the atmosphere. The third constituent of the dry atmosphere is argon, and its concentration is 0.93 percent by volume. The total volume occupied by these three gases is then 99.96 percent, leaving only 0.04 percent for other gases such as CO_2, ozone, neon, helium, methane, etc. Clearly the concentrations of these gases must be very small; they are measured in parts per million by volume. Notice that our estimate of the annual injection of CO_2 from vehicles was on the order of a part per million by mass per year. In order to convert from parts per million by mass to parts per million by volume, we use the laws of perfect gas mixtures (Dalton's law and the law of additive volumes), which state that the ratio of the number of kmols of a constituent, N_i, to the sum of the kmols of all the constituents in the mixture, N_m, is equal to the ratio of the volume of the constituent, V_i, to the total volume, V, that is,

$$\frac{N_i}{N_m} = \frac{V_i}{V}. \tag{1.4}$$

The number of kmols N_i and N_m are the masses in kg divided by the molecular weights, M_i and M_m, respectively, and so Equation (1.4) can be written

$$\frac{m_i}{M_i}\frac{M_m}{m_m} = \frac{V_i}{V}. \tag{1.5}$$

Here m_i and m_m are the masses of the particular constituent and mixture respectively. Thus, although oxygen occupies 20.95 percent of the atmosphere by volume, its ratio by mass is $20.95 \times 32/29 = 23.12$ percent. We have taken the molecular weight of the atmosphere to be 29, which is the weighted average of the molecular weights of N_2 and O_2, that is, M_m (atmosphere) $= 0.7808 \times 28 + 0.2095 \times 32 = 28.56$. We have rounded this up to 29 for our approximate calculation which neglects minor gases.

Let us return to the chemical makeup of the atmosphere. Although oxygen is in abundance, two other molecules that are essential to photosynthesis and thus life itself, CO_2 and H_2O, must exist only in very small concentrations because oxygen, nitrogen, and argon occupy 99.96 percent of the atmosphere by volume. Most of the water, is of course, in the sea, in ice, in lakes, and in rivers, and the amount in the atmosphere is highly variable from location to location; typical variation is from close to zero to over 3,000 parts per million of atmosphere by volume (which will now be abbreviated to ppmv). The carbon dioxide concentration is, today, close to 350 ppmv. Its value is carefully determined by means of sensors placed far away from cities (where the local concentration of CO_2 is higher), in such places as the South Pole or atop Mauna Loa, a mountain in Hawaii.

We are now in a position to determine whether automobile production of CO_2 is significant or not. Our estimate of the annual production by vehicles was 0.6 parts of CO_2 per million parts of atmospheric mass. From Equation (1.5) this is equivalent to $0.6 \times 29/44$ or approximately 0.4 ppmv per year. If we add in the production of CO_2 by coal and oil power plants, the burning of forests, etc., the amount of CO_2 added would be considerably higher; more than double this value. Increasing the amount of CO_2 in the atmosphere by one to two parts per million, from a present value of 350 ppmv, should be measurable over a number of years.

Figure 1.6 shows a graph of the carbon dioxide concentration in the atmosphere, in ppmv, from 1958 to the present at Mauna Loa. The saw tooth shape is due to the increase in removal of CO_2 in the northern-hemisphere summer by the large forests which are dormant in winter. The graph shows a clear upward trend of a little over 1 ppmv per year, and this is consistent with our calculations. Without having to resort to any other theories or hypotheses, we have determined that man-made (or anthropogenic) sources of CO_2 are enough to

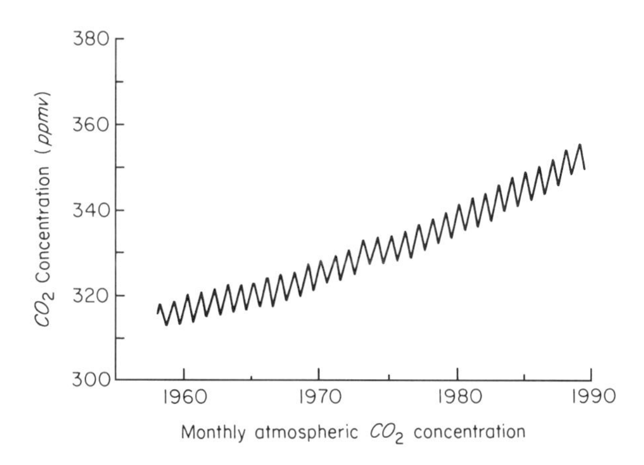

Figure 1.6 The concentration of atmospheric carbon dioxide, CO_2, in parts per million by volume (ppmv) measured at Mauna Loa in Hawaii. Measurements done at the South Pole look much the same. (After "Trends '93," CO_2 Analysis Center.)

explain the measured increase. What is the baseline value of CO_2? Ice core measurements suggest it has been around 300 ppmv for the preindustrial revolution period of civilization. We have increased the level by around 17 percent in less than 100 years.

I believe Figure 1.6 is one of the most important graphs of our time. It shows that we are, by means of the waste products of combustion, significantly changing the chemical makeup of the whole atmosphere.

1.4.3 The Engine and the Atmosphere – An Engineering System

In the old days, an automobile designer saw his or her design as an isolated object. The exhaust went into an essentially infinite sink; the small amounts of CO_2 added to the atmosphere were not measurable. Our analysis has shown that today the exhaust from all our cars is not going into an infinite sink but into a large but finite container. Because of this, the atmosphere now becomes part of the engine, and no car can be designed in isolation. Today, automobile design does not end at the tailpipe. Part of the engineer's concern must be for the atmosphere and for the total environment in general.

Although the increase in the CO_2 content of the atmosphere is measurable, and the fact that the increase is anthropogenic is clear, we now turn to the most difficult question of all. Does it matter that the CO_2 concentration of the atmosphere is rising by a few parts per million by volume per year?

Carbon dioxide is a harmless gas in small amounts; indoor concentrations can be well above 350 ppmv. Unlike carbon monoxide, CO, a gas that is often produced (in small amounts) by combustion, CO_2 is not toxic, nor is it very reactive. Its residence time in the atmosphere is about six years. We all know, of course, that it is a so-called greenhouse gas. Its molecular structure is such that it is transparent to the dominant short wavelengths of incoming solar radiation but opaque to the long wavelengths of outgoing radiation from the earth itself. The greenhouse effect will be studied in Chapter 5. Adding more CO_2 to the atmosphere is a little like putting on an extra layer of blankets. The more CO_2, the warmer the air becomes, and so to the possibility of global warming, melting of ice caps, catastrophic rises in sea level, etc. The mechanisms of the greenhouse effect, however, are much more complex, and we will show that the analogy to putting on extra blankets is weak. Moreover, water vapor is also a greenhouse gas, twice as effective as CO_2 in absorbing the outgoing long-wave radiation. A slight change in the atmospheric temperature will change the hydrological cycle; evaporation and condensation rates will change and so will the type and height of cloud cover. This will change vegetation growth rates as well as further complicate the warming process. More clouds, after all, could reflect the incoming sunlight and therefore allow for some global cooling. Or maybe they would absorb more? I will not address these questions and problems here; they will be discussed later on in the book. I will, however, give you my conclusion: there are strong chances that global warming is occurring, but we cannot be completely sure. The problem is too complicated for us to solve at present. There are unknown variables, and the atmospheric system is extremely complex: some have argued that it is the most complex physical system we know of. The above considerations do suggest, however, that the design of a modern transport system is a difficult task indeed, because we have shown that the millions of vehicles are affecting the makeup of the atmosphere, and therefore the atmosphere must be part of the design problem.

This indeterminate type of answer is typical of many important engineering problems. Because engineering in its very essence is social and environmental, very often, for problems that really matter, we cannot produce the clean, neat solutions that students would like to think exist. We can, of course, solve lots of little problems almost completely. For example we can design a shaft to transmit a certain amount of torque, or a tire that will provide the required traction on a wet surface. These are *components* of an overall *system* that is the automobile. The automobile is yet another component of an even larger system: the automobile fleet-atmosphere system. It is the systems, not the collection of components, that make engineering a complex and often indeterminate subject (Figure 1.7). The modern engineer must ask not only: How can we reduce the drag coefficient

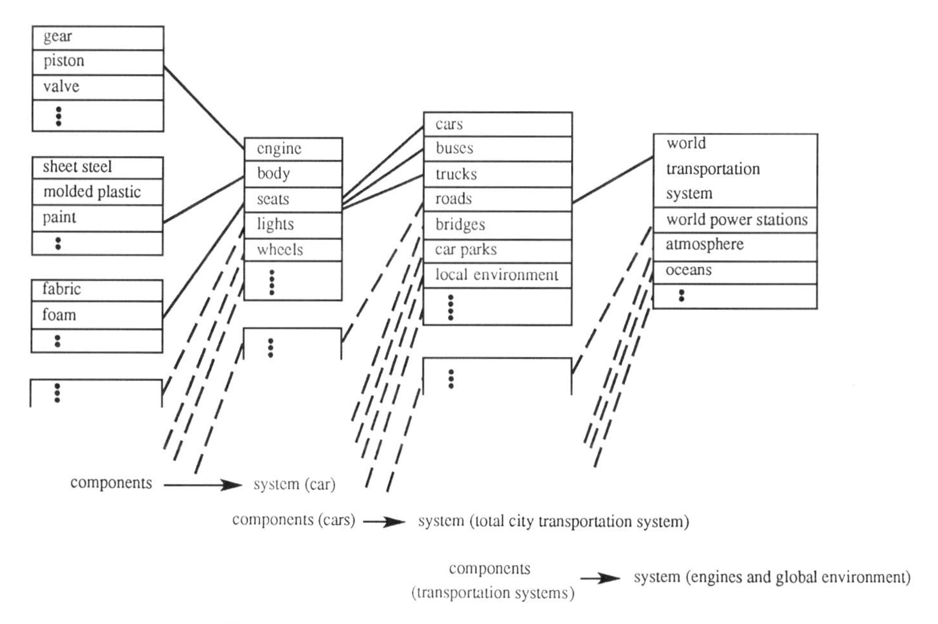

Figure 1.7 Components make up systems, groups of systems are components for (super) systems, and so on. As we move from the left to the right of this flowchart, societal, environmental, and ethical problems assume greater and greater importance.

of a car by a few percent, but: What type of transport should we design for the future? Should it be based on hydrocarbon fuels; should we be leaning toward mass transit systems or small, super-efficient, personal vehicles? Answers to questions such as these require immense technical skills, cutting across many disciplines.

1.4.4 Summary

We have shown that the combustion of a single gallon (3.785×10^{-3} m^3) of gasoline produces over 18 lb (8 kg) of CO_2, because the carbon atoms from the gas combine with the oxygen in the atmosphere. We have estimated that the hundreds of millions of gasoline engines produce around 3×10^{12} kg of CO_2 per year, and this is exhausted into an atmosphere whose mass we have estimated to be 5.2×10^{18} kg. Thus the CO_2 in the atmosphere is increasing by approximately 0.6 parts per million by mass or 0.4 parts per million by volume (ppmv) each year because of exhaust from vehicles alone. If we include the CO_2 produced by coal- and oil-fired power stations, forest burning, and domestic fuel burning, the increase is over 1 ppmv each year. This is a significant and measurable quantity because the total amount of CO_2 in the atmosphere is only 350 ppmv. In 1955 it was 310 ppmv, and the increase since then is due to combustion of fossil fuels. Because we are changing the nature of the atmosphere by fossil fuel combustion, I have argued that the atmosphere is now an essential part of automobile and power-plant design. Modern engineers must be as much aware of the overall system as of the components from which it is derived. The components of modern car design are not only cylinders and crankshafts, but the car itself, which is a component of the overall fleet, and the atmosphere, which is now recognized as a finite exhaust enclosure, not the infinite sink it appeared to be a few decades ago.

1.5 Outline of the Book

Because the role of the engineer is becoming broader, the traditional areas of study must be connected together at an earlier stage of the engineer's development. This does not mean, however, that the subject matter must be dealt with superficially. The types of questions that arise from this introductory chapter require answers that rely on a deep understanding of both the workings of the engine and of the atmosphere. In this book I will lay out the foundations of these subjects, so you will be able to better see how they are connected when you deal with them in detail in later years. My accent will be on basic principles, not superficiality. Chapters 2, 3, and 4 will deal with thermodynamics, fluid mechanics, and heat transfer. An engine is a thermodynamic system involving heat and work, which produce motion. The issue of engine efficiency is addressed in Chapter 2, which outlines the first and second laws of thermodynamics. Inside the cylinder of the engine is a gas or a liquid or a combination of both. These are fluids in complex motion. In Chapter 3 we will look at the foundations of fluid mechanics and how it relates to engine behavior. The heat is transferred from one part of the engine to another, or between the engine and the environment, by conduction, convection, and radiation. The laws governing heat transfer will be outlined in Chapter 4. In Chapter 5 we will apply the laws of thermodynamics, fluid mechanics, and heat transfer to the atmosphere, which has many aspects in common with the engine itself. We will return to engines in Chapter 6 and look at the sources of energy that we can use to drive them. In Chapter 7 we will look, as we have partly done here, at the engine–atmosphere system as a whole. Finally, we will return to the role of the modern engineer and discuss the complex issues he or she will face in the future.

The book is intended to be read as a whole. I would like you to read it more in the way you read a novel, rather than a text book. I have a story to tell, and just as with any good story, I hope you will learn more than the story itself. I hope you will be able to apply the approach I use here to other parts of your studies. The problems at the end of each chapter are also part of the story. They fill out concepts, and sometimes their answers will be used in subsequent chapters. Although many are rather short, and the mathematics required usually does not extend beyond arithmetic and algebra (only a handful requiring elementary calculus), they all require a fair amount of thought. Sometimes you will have to refer back to previous chapters for some empirical information or for a relationship that is needed. I ask the student to think broadly. After all, problems are meant to puzzle and provoke.

There are worked problems throughout the book, but unlike many of the texts you are familiar with, they are generally incorporated into the text rather than set aside at the end of sections. This is done to keep you busy as you read: it is essential to have a calculator at hand at all times to keep a close check on what is going on.

It is unlikely that the whole book can be read in one semester at the freshman level. When teaching it to incoming students I generally do all of Chapter 1 and most of Chapter 2 (but not too much on the second law). I then give a broad outline of fluid dynamics (Sections 3.1 and 3.2) and the Bernoulli equation (Section 3.3). I also discuss Sections 3.6 and 3.7 in very general terms. I then deal with the greenhouse effect (Section 5.2; with some reference to Section 4.3). At this point the student should be beginning to appreciate the approach used in the thermal–fluid sciences and why engines are important and how we go about understanding them. I then do Chapter 6, on energy sources, and finish with Chapter 7, which places the subject in an environmental and social context and relates the material back to Chapter 1. If the book is used for a second-semester freshman course (or for first-semester sophomore course), the pace can be faster and more material can be covered.

Finally, a note on units and accuracy. Whenever I do calculations for myself, I always convert all quantities to the Système International d'Unités (SI units). I always have a table of conversions close at hand. By using SI units (kilogram, kg; meter, m; second, s; degrees Kelvin, K; amount of substance, mol; and their derived units such as Joules, Newtons, Pascals, etc.), I know I will not introduce new sources of error, by having mixed units, into my calculations. I do all my research in SI units (and always set my examinations in them). However, in this subject of the engine and the atmosphere, there is a problem. In order to appreciate the issues, it is important to have a feeling for the magnitude of quantities. A fuel consumption of 25 miles per gallon means much more to you than a fuel consumption of 1.06×10^7 m/(m^3 of fuel) although they are both the same; the latter is in SI units. For this reason I will often use British or other units and put their SI equivalents in parentheses. I strongly recommend that you convert to SI units when doing a problem and then at the end, convert your answer back to the more familiar unit to see whether it makes sense. I also recommend that you carry with you a good set of conversion tables, both for now and for your later career as an engineer.

Now to the issue of accuracy of calculations. In Section 1.3 I made a distinction between precise and crude, or approximate, calculations and suggested that often great insight can be achieved by determining the solution of a problem to within a factor of two, or even to within an order of magnitude. This is not a book about precise calculations, and rarely will I quote values to more than one or two significant figures. This will cause a few minor problems. First, don't use this book to find out the values of such quantities as of the speed of light or the molar mass of water. I will use 3×10^8 m/s and 18 kg/kmol, not the more accurate values of 2.9979×10^8 m/s and 18.02 kg/kmol. (However, I have provided the precise values of these and other quantities in the end papers.) Second, sometimes answers will not be perfectly self-consistent; for example, using rounded values of molar mass for various substances may cause a slight violation of the conservation of mass in some calculations. You will have to judge for yourself whether you have made a real error or whether the lack of balance is due to rounding errors. Such judgment is part and parcel of the solution to many engineering problems. When precise answers are needed, the rounding errors are easily fixed.

1.6 Problems

1.1 In 1988 a satellite picture showed a cloud of smoke over the Amazon basin of three million km^2 ($\sim$1,000 miles on each side). From this and by other methods it was determined that one billion tons (10^{12} kg) of trees were burned that year. Assuming the reaction for biomass burning to be

$$CH_2O + O_2 + 3.76N_2 \rightarrow H_2O + CO_2 + 3.76N_2,$$

determine how much CO_2 (in kg) entered the atmosphere. If the mass of the atmosphere is 5.2×10^{18} kg and its mean molecular weight is 29, determine the volume ratio (in parts per million) increase in CO_2. How does this compare to the increase in CO_2 per year due to automobile emissions?

1.2 For the Kuwait oil fires after the Gulf War (1991) it was estimated that 6×10^6 barrels of oil per day were burned and entered the atmosphere. Assuming the oil to be approximated as C_6H_{14} (density; 660 kg/m^3), determine by how much the atmospheric CO_2 concentration (in parts per million by volume) increased after three months of burning. (one barrel = 0.1590 m^3)

1.3 The normal reaction of propane with air is

$$C_3H_8 + 5O_2 + 18.8N_2 \rightarrow 3CO_2 + 4H_2O + 18.8N_2.$$

Consider a reaction of propane when only four-fifths of the air required for a normal reaction is available. The left-hand side of the equation is known, but there is uncertainty about the products. It is thought that both CO and CO_2 may be produced. Thus the reaction may be written

$$C_3H_8 + 4O_2 + 15.04N_2 \rightarrow a\text{CO} + b\text{CO}_2 + 4H_2O + 15.04N_2.$$

(a) Determine a and b (hint: all atomic species must be conserved).
(b) If one kilogram of propane is burned in this reaction, how many kilograms of CO and CO_2 are produced?
(c) If one kilogram of propane is burned in the normal reaction, how much CO_2 is produced?

1.4 At present the total energy consumption of the United States is around 100×10^{18} J/year. What is the average rate of energy usage (in kW) for every American? If 80 percent of the energy is produced by fossil fuel, what is the average production of CO_2 (in kg) by all Americans per year? (Assume the fossil fuel is C_8H_{18}.) By how many parts per million by volume per year is the CO_2 atmospheric concentration increasing because of this? Note: The population of the United States is approximately 5 percent of the world population but its energy consumption is approximately one-third of the world energy consumption.

1.5 Compare the chemical energy of 1 oz (2.83×10^{-2} kg) of olive oil (about the amount on a salad) to the kinetic energy of a 1,000-kg automobile traveling at 60 mph (27 m/s). Assume the energy content of olive oil is 20×10^6 J/kg (about half that of gasoline). Is your answer surprising?

1.6 We have calculated the mass of the earth's atmosphere to be 5.2×10^{18} kg. It is always in motion relative to the earth, and therefore it has immense kinetic energy. Make an intelligent guess of the average speed of the air. Your guess will be accurate only to an order of magnitude. (Is it 1 mph, 10 mph? etc.) From this determine the total kinetic energy (in Joules) of the atmosphere. How does this compare to that of a 50 Megaton nuclear bomb, the largest ever made? (1 Megaton $= 10^6$ tons of TNT $= 4.2 \times 10^{15}$ J)

1.7 If we assume that the density of the atmosphere is constant, with a value of 1.2 kg/m^3 (the value at sea level), show that the atmosphere would have to be 8.5 km in depth to produce the observed sea level pressure of 10^5Pa. The density of the atmosphere in fact decreases with height (see Chapter 5), and thus it extends to a much greater height then 8.5 km.

1.8 Collect articles from journals such as the *New York Times, Newsweek, Time*, etc. Is there a consensus on whether global warming is occurring? Do particular opinions come from groups with vested interests (e.g., automobile manufacturers, environmental groups)? Are scientists in agreement about whether global warming is occurring or not? Do they have a vested interest? Survey your fellow students. What percentage think global warming is happening? Ask them how they arrived at their opinion. Do they have vested interests?

Symbols

A	area	m^2
F	force	N
g	acceleration due to gravity	m/s^2
M	molar mass or molecular weight	kg/(kmol)
m	mass	kg
m_a	mass of atmosphere	kg
N	number of moles	kmol
p	pressure	Pa
V	volume	m^3
η_{th}	thermal efficiency	
ρ	density	kg/m^3

2

Thermodynamics

2.1 What Is the Most Efficient Engine?

The billions of engines that are operating throughout the world use billions of gallons (or liters) of fuel. We have shown in Chapter 1 that the combustion process significantly adds to the carbon dioxide content of the atmosphere and this may cause *global* climate change. We also know that combustion causes *local* pollution effects, such as the photochemical smog of Los Angeles and Mexico City, and that the world supplies of gasoline are unevenly distributed, leading to international tensions and sometimes conflict, such as the Gulf War of 1991. Increasing engine efficiency would at least reduce these problems (Figure 2.1). In this and the next two chapters, I will outline the fundamentals of the thermal–fluid areas of engineering. These are needed to analyze engines, although as you will see, their application is much greater than this. We begin, in this chapter, by asking the question: What is the maximum thermal efficiency that an engine can have? Can it be unity, so that all the heat input is converted into work (Equation (1.1))? We will show that if this were so there would in principle be no energy problem at all.

To determine the efficiency of an engine, we must define and study the relationship between heat and work. This we will do in Section 2.4 after we have developed the necessary preliminaries. The notion of heat implies that there will be temperature differences. In subjects that you have previously studied such as Newtonian mechanics and electromagnetism, it is assumed that everything is at a constant temperature. In thermodynamics, the effects of temperature differences are of central importance.

Engines convert heat into work to drive automobiles, power generators, and the like. In a similar way the atmosphere converts heat into motion, producing approximately regular flows such as the jet stream high in the atmosphere as well as the chaotic, gusty motion at ground level that we are all familiar with. Work is important here too, because the motion would cease quickly due to

Figure 2.1 Our quest for fuel can cause war and environmental devastation. Increased engine efficiency would help to alleviate these problems. (Photo by W. F. Hunt, Jr., courtesy of *Journal of the Air and Waste Management Association.*)

friction if there was no work interaction between the various air masses. In this chapter we will develop the laws of thermodynamics, by using the engine as an example of a thermodynamic system. We could equally well have used other systems such as the atmosphere or the human body because each of these systems converts heat into work (Figure 2.2). The laws of thermodynamics are extremely general, applying to everything, large or small, living or inanimate, irrespective of chemical, physical, or biological makeup. They were first enunciated in the early part of the nineteenth century by engineers who were trying to maximize the efficiency of heat engines. Little did they realize that the laws they were to discover would have ramifications in all areas of science and technology.

There are two fundamental laws of thermodynamics from which the rest of the subject is deduced. A law is a generalization from experience and is not directly provable. An important objective of science is to establish laws that are universal, that is, laws that hold under all possible conditions. The laws of thermodynamics are considered to be the most general that have yet been formulated. No deviations have been found. Often, and particularly in thermodynamics, we design "thought experiments" to test the validity of laws. We ask, what would the world look like if these laws did not hold? We will see that if the first and second laws of thermodynamics did not hold, the world would be fundamentally different.

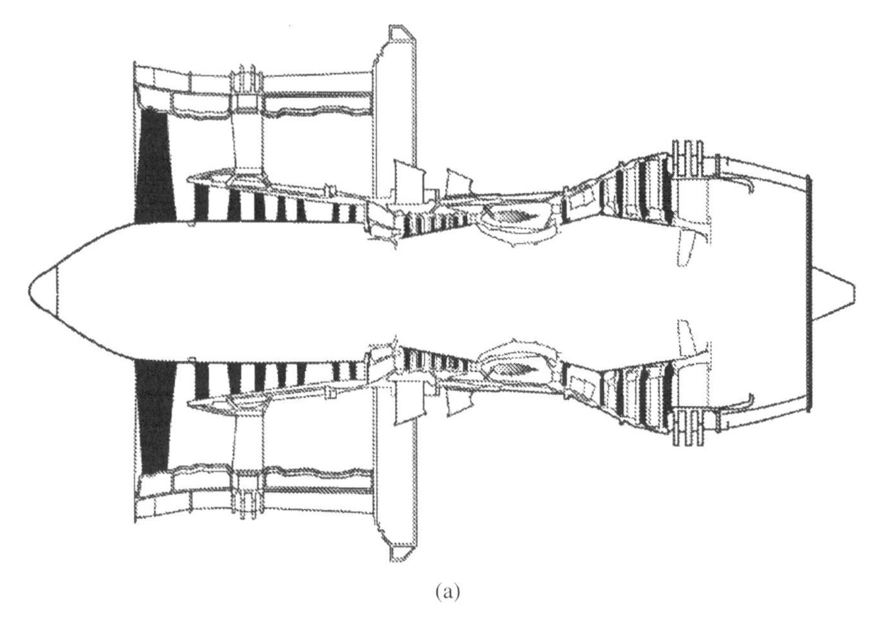

(a)

(b)

Figure 2.2 A jet engine and the earth's atmosphere are two examples of thermodynamic systems. In both cases temperature differences produce heat interactions that produce motion and work. (Part (a), sketch of a Rolls-Royce turbofan engine courtesy of Professor E. Ressler, Cornell University. Part (b), NASA.)

The first law of thermodynamics is a statement of the conservation of energy, and the second law states that it is impossible to have a heat engine that has an efficiency of unity. Telling you this, however, does not help much in the design of an engine. In order to make these laws operational, concepts such as **heat**, **work**, **energy**, **temperature**, **heat engine**, and **thermodynamic system** must be clearly defined. The early part of this chapter will be occupied with such definitions. Because the laws of thermodynamics are general, we will be able to use them without modification when we study the atmosphere in Chapter 5.

2.2 Thermodynamic Systems, Thermal Equilibrium, and Temperature

When we look at an engineering system, such as the engine in Figure 1.2, we are immediately struck by its complexity. We may ask: How do we even begin to analyze such a complex structure in order to determine its efficiency and how much work it will produce? Up until now, your scientific education has largely been focused on analyzing small parts of complex systems, without much consideration of the whole. For example, you have been more involved with the structure of individual atoms and molecules than how they aggregate to make up liquids or solids, let alone complex machines. In order to make the task of analyzing engines and other complex systems manageable, engineers often forego looking at the details and view the system as a whole. Instead of trying to describe the precise motion of the gas inside the cylinder or the action of the high-voltage spark when ignition takes place, they ask broad questions of the form: What does the engine do? What input does it need to do it?

To answer these questions we must carefully define what a system is. We formally define a **thermodynamic system** as a three-dimensional region, bounded by a surface. The region may have rigid or deformable boundaries, and its volume or mass may change or remain fixed. The task of thermodynamic analysis is then to study what crosses the system **boundary**, that is, the interface between the system and the surroundings. We will show that work, heat, or mass may flow across the system boundaries, thereby changing the characteristics of the system. If no mass crosses the system boundaries, the system is referred to as a **closed system**. Such systems will be studied in this chapter. In Chapter 3, we will study **open systems**, those for which matter (as well as heat and work) crosses their boundaries.

The thermodynamic system itself is defined in terms of its **properties**, which are the observable characteristics of the system, such as mass, volume, density, pressure, energy, and any other relevant aspect that is necessary to specify the system at a particular time. The **state** of the system is the totality of its properties. For example, at a particular moment we may describe the state of a system consisting of gas inside a cylinder in terms of its pressure, volume, and temperature. Or if the system is a falling rock, its state would be described by its mass and kinetic and potential energies. A **process** is any transformation of a system from one state to another.

Notice that our formal definition of a thermodynamic system does not provide any information on how to determine where the system's boundaries are to be drawn. Knowing how to separate a system from its **surroundings**, that is, all of the universe not included in the system, is a central task in engineering analysis.

The first step in determining what is the system and what is the surroundings is to ask the question: What do I want to know? (Figure 2.3) If, for example, we wish to know the overall efficiency of an automobile, we would take the whole car, including wheels, as the system. Its boundaries would be the paint

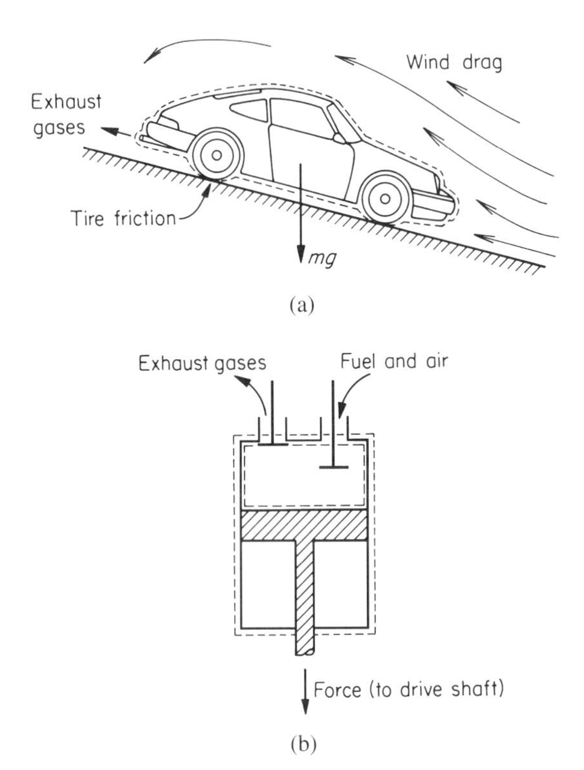

Figure 2.3 Clearly defining the system and its boundary is of central importance in doing problems in thermodynamics. In Figure 2.3(a) the whole car is the system (with the dashed line showing its boundary). In Figure 2.3(b) we consider the piston-cylinder only. Here I have shown two possible choices for the thermodynamic system: the whole piston-cylinder assembly (the outside dashed line) and the gas inside the cylinder (the inside dashed line). The choice of system depends on the type of question being addressed. In this chapter we will be focusing particularly on what is occurring inside the cylinder, so the inside choice will be more appropriate.

and tires, etc. We would determine how much gasoline would cross the system boundaries in order to produce a certain amount of power to overcome air and road resistance at its boundaries, and the gravitation force acting on the system. If, however, we were concerned with the engine itself, our system might be the piston and cylinder. Choosing the system is often quite straightforward, but sometimes it requires judgment and creativity. It is the most important part of an engineering analysis because if it is wisely chosen, the rest of the problem is relatively straightforward. On the other hand, a poorly defined system may make the solution difficult, and sometimes impossible. Note that the system boundary can consist of a real surface, such as the inside of a cylinder, or it can be imagined. For example, in Chapter 3 we will sometimes define an arbitrary region of fluid as the system. The examples in this and the following chapters will provide you with experience in defining the system, but as previously mentioned, the most important step is to ask: What do I want to know?

Once we have defined the system, we ask whether it is in **equilibrium** or not. A system is in equilibrium if there is no tendency toward change (providing, of course, that there are no changes in the surroundings). Take a ball rolling in a bowl, an example to be dealt with in detail in Section 2.4. While the ball is rolling, this system (consisting of the ball and the bowl) is changing all the time and is therefore not in equilibrium. Once the ball has come to rest at the bottom of the bowl, the system will be in dynamic equilibrium because there will be no tendency for the ball to change position unless the system is acted on from outside.

However, because of friction, there will be a lack of complete equilibrium even when the ball comes to rest at the bottom of the bowl. The motion of the ball will cause its own material and the material of the bowl to change in some way (for example, they will expand slightly if they are made of copper). If the system is open to the surroundings (i.e., if it is not inside an insulated container), after a time the metal will contract to its initial density. This contraction will occur without the imposition of any external force. We say that the system is now in **thermal equilibrium** with its surroundings, in distinction to the dynamic equilibrium that was achieved when the ball came to rest. In everyday language we would say the system and its surroundings are at the same temperature. (If, on the other hand, the ball and the bowl were in an insulated box, the contents would rise to a higher temperature as the ball came to rest.) Indeed, in thermodynamics we define **temperature** as the property that determines whether a system and surroundings are in thermal equilibrium. This is the basis of thermometry: the mercury in a thermometer expands (or contracts) when placed in surroundings that it is not in thermal equilibrium with. The expansion stops when thermal equilibrium is attained, and suitably calibrated the thermometer can be used to define a temperature scale. In general, changes in pressure, conductivity, elasticity, and other properties may occur while thermal equilibrium is being attained. The essential point is that when thermal equilibrium has occurred there will be no change in any property of the system unless there is a change in the surroundings.

2.3 Engineering Cycles and Heat Engines

Systems tend to undergo cyclical motion: wheels rotate, pistons go up and down, planets rotate and also orbit the sun. More complex systems, such as the earth's atmosphere, also go through cycles. And there are cycles in biology, such as the motion of the heart and lungs.

The cyclical nature of systems allows us to employ simplifying concepts in their analysis. A **cycle** is defined as a process that a system undergoes such that its initial state is identical to its final state. For example, for the engine of Figure 1.5, the piston returns to its original position after going through a cycle. Moreover, the temperature and pressure of the gas inside the cylinder also return to the same values they had before the cycle commenced. At any stage of its cycle, the system is completely defined in terms of its properties, and for the piston engine, the relevant properties are the temperature, pressure, and volume. Clearly the properties of a system are the same for its initial and final states if it has undergone a cycle.

A cycle implies a path. For example, a rotating wheel goes through a circular path. The notion of a **path** can be generalized to defining the state of the system for every moment of its change. For the piston cylinder, the path is the value of the pressure, temperature, and volume at every stage of the cycle; for the heart, it is the oxygen level, muscle tension, etc. A system changing between the same end states may go through different paths. For example, Figure 2.4 shows two paths for a gas being expanded in a cylinder. States 1 and 2 are the same, but the paths are different. We will show that the amount of work an engine does, as well as its efficiency, is dependent on its path. Note that to define a path, the

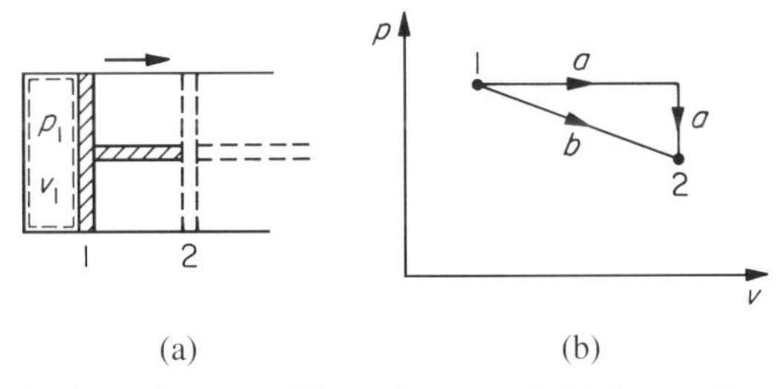

Figure 2.4 A piston is moved from the same initial state (p_1 and v_1) to the same end or final state (p_2 and v_2), but the paths for the system are different. Here v is the specific volume (m^3/kg) and p is the pressure (Pa). The system boundary is shown by the dashed line. It changes as the piston moves out.

state of the system must be able to be specified at all times. This implies that the different parts of the system cannot be too far from equilibrium with respect to each other as the system changes state. For most engineering problems this is a reasonable assumption. Thus for a gas in a cylinder, we will assume that at any given moment the pressure and temperature are constant throughout the system, although in fact there will be some unevenness of these properties.

Finally, we must define what we mean by an engine. We have so far used the word in a loose way, to mean any device that does work. In our analysis to follow, we will define a **heat engine** as a system that operates in a cycle and produces work, with only heat (no mass) crossing its boundaries. The usefulness of this restricted type of engine will become apparent as the analysis proceeds.

With these preliminaries, we will now turn to the application of the first law of thermodynamics to engineering systems. Because this law is universal, we will begin by looking at a very simple system that we are all familiar with. This will allow us to make our concepts of system, cycle, path, and state operational and will allow us to define heat and work. We will then apply these concepts to the analysis of an engine.

2.4 Energy, Work, and Heat: The First Law of Thermodynamics

Consider a ball rolling in a bowl, in a container, as shown in Figure 2.5. Let the ball start its motion at position 1. (How it got there originally does not concern us for the moment.) It will roll down the bowl and not quite reach the same height on the opposite side. After a number of ups and downs, it will stop at the lowest point of the bowl, at position 2.

What is the thermodynamics of this process? In order to address this question, we must first ask: What do we want to know? Asking this question will enable us to determine the thermodynamic system, the first and most important step in our analysis. If we want to know how the ball itself changes its energy or temperature, it would be sensible to make the ball the system and everything else the surroundings. On the other hand, recognizing that the rolling ball may affect the temperature of the bowl, and the surrounding air, we may want to look at the overall effects of its motion. Here we would take the ball, bowl,

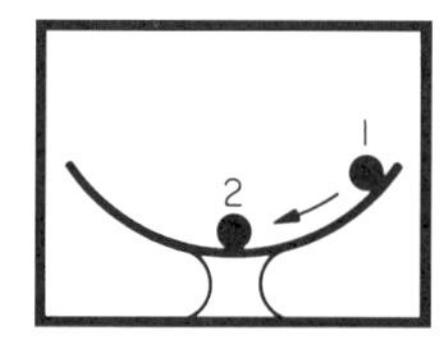

Figure 2.5 A ball in a bowl. This thermodynamic system is isolated from the surroundings by means of an insulated box, which is the system boundary.

and surrounding air as the system and the box as the system boundary. It is this second situation that we will analyze here because it has some of the complexity akin to engineering systems.

To begin with I will place a special constraint on the system comprising the ball, bowl, and air: the box is insulated. That is, nothing that occurs in the box affects the outside world (or vice versa). For example, if the temperature should change inside the box, there will be no temperature change outside. For this situation the box is the boundary of an **isolated system** that comprises the ball, bowl, and enclosed air. We will lift this constraint after we have analyzed this special case.

Let us pretend for the moment there is no friction. The ball will now roll forever, having its maximum velocity at the bottom and zero velocity at the top of the bowl where its direction of motion changes. The kinetic energy of the ball is $\frac{1}{2}mV^2$, where m and V are the ball's mass and velocity respectively, and the potential energy is mgz, where g is the acceleration due to gravity and z is the height of the ball. Note that in order to define potential energy we must have a reference for z. It could, for example, be sea level. Here we will take it as the bottom of the bowl. For this situation, the conservation of energy, or the first law of thermodynamics for our system, is

$$mgz + \frac{1}{2}mV^2 = \text{constant} \qquad \text{or} \qquad \Delta\left(mgz + \frac{1}{2}mV^2\right) = 0. \tag{2.1}$$

As the ball roles down the bowl, z decreases and V increases such that Equations (2.1) hold everywhere. The Δ sign in the second Equation (2.1) is short for "the change in." Equations (2.1) can also be written in the form

$$\left(mgz + \frac{1}{2}mV^2\right)_a = \left(mgz + \frac{1}{2}mV^2\right)_b$$

or

$$mg(z_b - z_a) + \frac{1}{2}m(V_b^2 - V_a^2) = 0, \tag{2.2}$$

where a and b are any two positions of the bowl. Notice that for our frictionless example, we have had to worry only about the ball itself, because nothing is happening to the rest of the system. (In fact there will be a small amount of air motion induced by the moving ball, but we will neglect this for the moment.)

Now, we know that the ball will not go up and down for ever. Friction will cause it to stop at the bottom after some time. Place a thermometer in the box (the isolated system) and after the ball is at rest at the bottom and everything has settled down or come to thermal equilibrium within the box, the temperature of the contents of the box will have risen a slight amount. If you could observe the interface between the ball and the bowl with an extra-ordinarily powerful microscope, you would see that the surface of the ball, as it speeds down the bowl, causes the atoms of the ball and bowl material to vibrate faster. In time these excited atoms excite their neighbors such that the excitement spreads throughout the bowl and ball and the air in the box also. The atoms now have extra energy. Because it is random, we call it **internal energy**, but it really is random kinetic energy at the atomic scale. Note that if it were not random, that is, if all the atoms were excited in one direction, the whole bowl would move. This does not occur! The increase in internal energy is manifested as a rise in temperature.

For this situation, the first law states

$$U + mgz + \frac{1}{2}mV^2 = \text{constant}, \tag{2.3}$$

where U is the internal energy of the system. At this stage we cannot determine U from first principles, but we can evaluate its change (if we know the change in potential and kinetic energy) from the equation

$$\Delta\left(U + mgz + \frac{1}{2}mV^2\right) = 0. \tag{2.4}$$

In Equations (2.3) and (2.4) the kinetic energy term has been retained for generality, but you should realize that for our example, $\Delta(\frac{1}{2}mV^2) = 0$ and that the decrease in potential energy is equal to the increase in internal energy after the ball has come to rest at the bottom of the bowl. Thus, between state 1, the initial condition at the top of the bowl, and state 2, its final position at the bottom, the first law is

$$\Delta(U + mgz) = 0$$

or

$$U_2 - U_1 + mg(z_2 - z_1) = 0. \tag{2.5}$$

Note that the initial state is always subtracted from the final state to find the change of a property.

Equations (2.3) and (2.4) state that for a process, the sum of the internal, potential, and kinetic energies of an isolated system (a system that does not interact in any way with the surroundings) is constant. This is a statement of the **first law of thermodynamics**. While the kinetic and potential energy refers to the macroscopic aspects of the system (bulk motion and vertical displacements), internal energy deals with the microscopic aspects. We have mentioned that the internal energy can be changed by means of friction. We will soon show that it also can be changed by compressing a perfect gas, even under frictionless conditions. In both cases there is a rise in the temperature of the system. Internal energy is also stored in the bonding of molecules and atoms and can be released when there is a chemical or nuclear reaction. Or it can be stored in a spring and be released (in the form of kinetic energy) as the spring unwinds. The essential point as far as the first law is concerned is that there are only three forms of energy – kinetic, potential, and internal – and the sum of these remains constant for an isolated system undergoing any process.

So far we have studied the thermodynamics of an isolated system, going from one state to another. Although its total energy has remained constant, it has changed in form; for our example of the ball in the bowl, the potential energy of the initial state has all been converted into internal energy at its end state. Yet our purpose is to understand engines, which work in cycles. Thus we will further explore our ball-in-the-bowl example, allowing it to go through a full cycle.

In order to complete the cycle we must return the system to its initial state, that is, to return the ball to its position at the top right-hand side of the bowl and to reduce the temperature of the system to what is was before the ball started rolling, such that an observer would not know that anything had occurred within the system. How can we do this? Because the ball will not roll up the side by itself (and if it did, the temperature would increase even further because of friction), it appears that the only way we can achieve the initial state is to interfere with the system from the outside, or, in thermodynamic terms, have an interaction across the system boundaries. First, in order to reduce the temperature of the system, we must have a heat interaction; we must open one side and let heat flow from the system to the surroundings until the temperature of the system is the same as it was before the ball started rolling (Figure 2.6(a)). **Heat is defined as an interaction across the boundaries of a system that occurs by virtue of a temperature difference.** It is an interaction between a system and

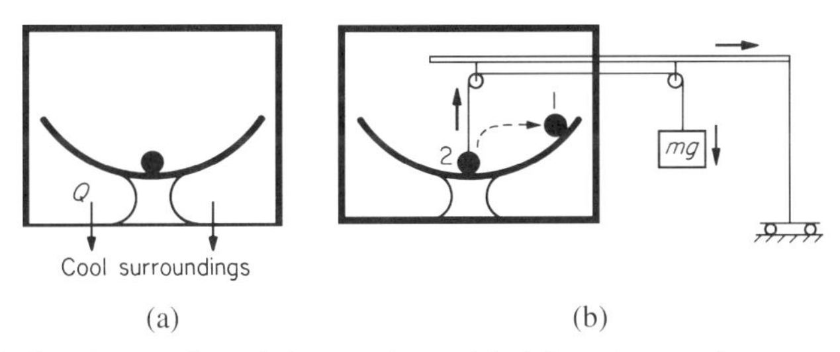

Figure 2.6 Heat and work interactions. (a) A heat interaction across the system boundary. The system is at a higher temperature than the surroundings as a result of the frictional motion of the ball. The insulation at the bottom of the box is removed, and there is a heat interaction from the system to the surroundings. (b) A work interaction across the system boundary. The ball is lifted from the bottom of the bowl to its original height. Thus work is done on the system. (In order to get the ball back to the side of the bowl [its original position] after the weight has been lifted, it must be moved laterally to the right [by sliding the whole pulley mechanism]. Because the height of the ball does not change, no further work is done on the system.) For this example the work and heat interactions have restored the system to its initial state.

its surroundings. In thermodynamics there is no such thing as a heat interaction within a system. A heat interaction is a transfer of energy, and in this case it will cause the internal energy of the system to decrease. The symbol for a heat transfer is Q. Our sign convention will be that if heat is transferred to the system, Q will be positive, and if it is transferred away, Q will be negative. As for energy, its unit is Joules.

Next, we must return the ball to its initial state. This is done by a work interaction across the system boundaries. Either with our arm, or by means of a pulley (attached to a weight external to the system), we lift the ball back to its original position (we do this slowly and keep the ball away from the bowl in order to avoid friction, which would cause the system temperature to rise again, Figure 2.6(b)). Work, like heat, is an interaction across the boundary of a system. In thermodynamics there is no such thing as a work interaction within a system. **Work is defined as the interaction between a system and its surroundings when the sole effect external to the system could be the lifting (or lowering) of a weight**. As for heat and energy, its unit is Joules. The symbol for a work interaction is W. Our sign convention will be that if work is transferred to the system (equivalent to lifting a weight in the system or lowering one external to it), the work interaction is positive. Work transferred to the surroundings is negative.*

After the heat and work interactions are complete we close up the system. Its internal energy, potential energy (and kinetic energy, which is zero) will have been returned to their initial values, that is, the state of the system will be exactly what it was before we started. It will have gone through a cycle.

The process of cooling the system and returning the ball from the bottom of the bowl to its initial state at the top may be written symbolically as follows. Reducing the temperature to what it was before the ball rolled down the bowl requires a heat interaction. The first law for this process is

$$Q = \Delta U, \tag{2.6}$$

where Q is the heat transferred (Joules) and ΔU is the change in internal energy of the system. The sign of Q will be negative because heat is being transferred from the system to the surroundings. Thus ΔU will also be negative.

Lifting the ball (without friction) to the top of the bowl requires a work interaction:

$$W = mg\Delta z. \tag{2.7}$$

* You should be aware that the opposite sign convention is also used in some books: Work is taken to be positive if it is done by the system (i.e., if it lifts a weight external to it). There is no "correct" sign convention just as there is no "correct" traffic convention concerning the side of the road we drive on. But beware if you use the U.S. convention and drive on the right-hand side of the road in England! And beware if you mix your conventions in doing thermodynamics problems. In both cases there will be a disaster.

Here W is the work interaction (Joules). Because here Δz ($= z_1 - z_2$, Figure 2.6) is positive, so too will be the work interaction, and this is in accordance with our sign convection.

The first law for the combined process of cooling the box and lifting the ball is the sum the heat and the work interactions (Equations (2.6) and (2.7)):

$$Q + W = \Delta(U + mgz). \tag{2.8}$$

Now, for the ball rolling down the bowl all the potential energy was converted into internal energy (Equation (2.5)). Here by means of a work and heat interaction we have restored the internal and potential energies to their initial values. Therefore the right-hand side of Equation (2.8) must be zero, the increase in potential energy ($mg(z_1 - z_2)$) balancing the decrease in internal energy. This implies that

$$Q + W = 0. \tag{2.9}$$

Equation (2.9) is very important. It is a statement of the first law of thermodynamics for a cycle. Although derived for the simple ball-in-the-bowl example, it must hold for any cycle because by definition, for a cycle the total change of energy $\Delta(U + \frac{1}{2}mV^2 + mgz)$ must be zero. (The initial and final states are the same.) Generally there will be more than one heat and work interaction. In this case their algebraic sum must be zero too.

Note that the signs will take care of themselves. For example, if 10 kJ of work is done lifting the ball, W is positive because it is done on the system. Then Equation (2.9) shows that $Q = -10$ kJ. It is negative, indicating that there is a transfer of heat from the system to the surroundings, as there must be to cool the system, returning it to its initial state.

In fact we have analyzed a friction machine in which work is converted into heat. We may run this machine through cycle after cycle, lifting the ball (doing work on the system), letting the ball roll down and thereby increasing the internal energy of the system, and then letting the system cool and thereby heating the surroundings.

Of course we require the opposite type of process to occur in an engine: We require heat to be converted into work. Intuition tells us that the machine we have just described cannot be run backwards, that is, that by heating the system the ball will not roll up to the top of the bowl (making itself available to lift a weight) although the first law for the cycle (Equation (2.9)) does not preclude this possibility because all it states is that the heat and work interactions must be equal in magnitude for the cycle. It makes no distinction between heat and work interactions. Why our friction machine cannot be run backwards will be explained when we study the second law of thermodynamics.

Clearly, a machine that converts heat into work must be more complex than the one just analyzed. We will describe such a machine after we have summarized the first law.

2.5 Summary of the First Law

The first law is an energy budget: The energy of a closed system can be increased or decreased only by work or heat interactions, otherwise it remains constant, although conversions from one kind to another can occur within the system. It is a little like a bank. Money can enter or leave (positive or negative heat and work interactions), thereby changing the total assets (energy). Inside the bank the type of assets may be converted from one form to another (e.g., from stocks to bonds) but the total assets remain the same. This is equivalent to changing the form of energy within the system from potential to internal, for example. Notice that although bonds may be converted back to stocks, internal energy cannot be converted back to potential energy without interactions across the system boundary. This is a second-law issue and will be discussed in Section 2.10.

More formally, for a closed system, the sum of the work and heat interactions is equal to the total change in energy of the system, ΔE, or

$$\sum Q + \sum W = \Delta E = \Delta\left(U + mgz + \frac{1}{2}mV^2\right), \tag{2.10}$$

where the summations mean that there can be more than one heat or work interaction. Often, (as we will see for a piston engine) Δmgz and $\Delta\frac{1}{2}mV^2$ are zero, and then Equation (2.10) becomes

$$\sum Q + \sum W = \Delta U. \tag{2.11}$$

In doing calculations it is extremely important to keep proper track of signs. The Δ symbol means "a change in." Its sign will be correctly determined by always determining the change as the difference between the final and initial states.

If the system goes through a *cycle*, returning to precisely the same *state* it started from, then the first law becomes

$$\Delta\left(U + mgz + \frac{1}{2}mV^2\right) = 0\,[\text{cycle}]. \tag{2.12}$$

For a cycle, it follows from Equation (2.10) that

$$\sum Q + \sum W = 0\,[\text{cycle}]. \tag{2.13}$$

2.6 Alternative Forms of the First Law

When we study gases in cylinders we will be interested in changes per unit mass. To do this, we divide each side of the first law by the mass, m, converting **extensive** properties to **intensive** properties. Equation (2.11) becomes

$$\sum q + \sum w = \Delta u, \tag{2.14}$$

where $u \equiv U/m$ etc. If the system is isolated, $\sum q = \sum w = 0$ and

$$\Delta u = 0 \,[\text{isolated system}] \tag{2.15}$$

or more generally

$$\Delta\left(u + gz + \frac{1}{2}V^2\right) = 0 \,[\text{isolated system}]. \tag{2.16}$$

For a cycle, where the end state is the same as the initial state,

$$\sum q + \sum w = 0 \,[\text{cycle}]. \tag{2.17}$$

Finally, sometimes we will want to discuss small, or *differential*, changes of a system. For this we will write the first law as

$$\delta q + \delta w = du. \tag{2.18}$$

This is to be read as "the sum of small amounts of the heat and work transfers results in a small change in internal energy of the system." The reason for using the δ signs for w and q and the normal differential sign for u is to distinguish between an interaction across a system boundary (the δ sign) and a change of a property within the system (the d sign). Below we will provide a further reason for this distinction.

2.7 Application of the First Law to a Piston Engine

Our example of the ball in the bowl (Section 2.4) illustrated the concepts of energy, work, and heat interactions, and thermodynamic cycles. We used the first law to describe a machine that went through a cycle, converting its work input into heat output. We will now describe a much more useful machine, one that can convert heat input into work output: a heat engine. We will choose for our example the piston-cylinder, briefly discussed in Chapter 1. This is the most common way of converting heat into work, although there are many other possibilities such as heating and cooling elastic materials (such as steel springs or rubber bands) that could produce cyclical motion because of their expansion and contraction. The first law of thermodynamics holds for all matter, be it a solid, liquid, gas, or a combination of these states.

Because we are interested in determining the net work output from a given heat input, we will choose as our thermodynamic system the gas inside the piston-cylinder, with the walls of the piston and cylinder as the system boundary (Figure 2.7). We will assume that the system is closed (no mass crosses its boundaries). There will of course be work and heat interactions: The system is closed but not isolated. Note that the piston-cylinder system changes volume as

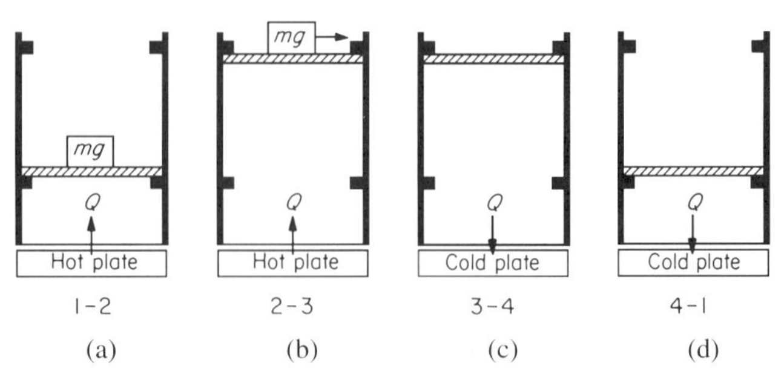

Figure 2.7 A heat engine. The gas inside the cylinder is the system. Both heat and work interactions occur across the boundaries. The cycle consists of four stages shown on the diagram.

the gas expands (in contrast to the ball-in-the-bowl system, for which the volume of the box was fixed).

Just as for the ball-bowl-box system, we must say something about the nature of the boundaries of the cylinder. We will be heating the gas inside the cylinder by means of a hot plate. We will assume that when a hot plate is applied to the base of the cylinder, heat is perfectly conducted to the gas. Further, we will assume that the heat affects only the gas, that is, none is lost to heating the walls of the cylinder, which we assumed to be perfect insulators. In practice some heat will be lost to the walls. Finally, the only significant energy change in the system will be the change of its internal energy, U. As the gas expands, its center of mass will change slightly and there will be motion of the gas inside. However, the change in potential and kinetic energy will be slight compared to changes in internal energy.

For simplicity, the cylinder will be filled with a **perfect** or **ideal gas**, obeying the law

$$pv = (R_u/M)T = RT, \tag{2.19}$$

where v (the reciprocal of the density, ρ) is the specific volume (m^3/kg), R_u is the universal gas constant (8.314×10^3 J/(kmol K)), M is the molar mass or molecular weight, and T is the absolute temperature in degrees Kelvin (K). The gas constant, $R \equiv R_u/M$. Thus for air, $R = 8.314 \times 10^3/29 = 2.87 \times 10^2$ (J/kg K). The analysis that we are about to do could be done for other substances filling the cylinder such as a combination of water and steam undergoing phase changes, or an even more complicated substance such as peanut butter and jelly, so long as it expands on heating. For either of these substances the first law will apply but the equation relating pressure and volume, the **equation of state**, will be much more complicated than Equation (2.19). Often the relation will exist only in tabular form because no simple expression can be found.

A hot plate is placed at the bottom of the cylinder, as shown in Figure 2.7. We wish to do work, to lift a weight, shown here sitting on top of the piston. (Of

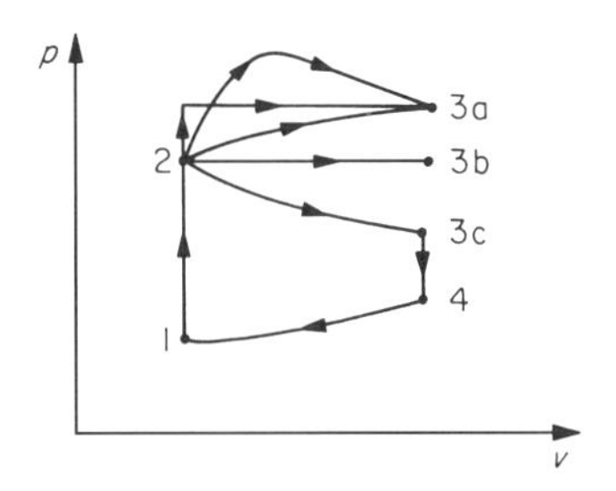

Figure 2.8 Possible p-v diagrams for the heat engine of Figure 2.7.

course in practice, the piston would be coupled to a device such as the slider-crank mechanism of Figure 1.5. The lifting of a weight is the same as far as the system is concerned.) At first, assume the heat transfer from the hot plate merely allows the gas pressure to increase, without any movement of the piston. This process is shown on the p-v diagram, for the air inside the cylinder, in Figure 2.8. The constant volume process is a vertical line from 1 to 2. When the pressure inside the cylinder is infinitesimally greater than the pressure of the piston and the weight, the piston will begin to rise, lifting the weight and doing work. How does the specific volume change with pressure? We do not, without making further assumptions, know the answer to this question, except that v must increase because the volume is increasing (and the mass is constant). There are an infinite number of paths, a few of which are shown in Figure 2.8. For path 2–3a, the heat transfer is so great that although v increases, so does p (and therefore so does T). On the other hand, for path 2–3c the pressure decreases, whereas for path 2–3b it remains constant. Notice, as we discussed in Section 2.3, that the nature of the path between two identical end states can also be different: a few possibilities are shown for the path 2–3a. A basic issue in engine design is What is the best path to take, to produce the most efficient engine? For us this is the central issue of this chapter.

Let us assume that for the particular geometry of this engine and the amount of the heat transferred, the path taken for the expansion is 2 to 3c. Once the piston has shifted to its outermost position, the weight will be slid off; work will have been done. Then we wish to return the piston to its original state, so another weight can be lifted up. After all, lifting one weight is equivalent to one-half of a rotation of a slider-crank mechanism, and to move a car or generate electrical power we must have continuous rotation. Returning the piston to its original state requires a decrease in v, but the pressure inside the cylinder may be too high (note that the weight has been removed), so at first a cold plate will have to be placed near the cylinder (Figure 2.7c) so that a heat transfer will take place reducing T and p. We will assume that v remains constant. This is path 3c–4. When the pressure diminishes sufficiently (at 4), the piston will move down, reducing v. The cold plate will remove just enough heat so that the system returns to its initial state, 1. Here too there are an infinite number of paths from 4 to 1, although only one has been shown. For this process work is done on the system to return it to its original state.

We will now show how to determine the work done and the efficiency of the cycle 1–2–3c–4–1 shown in Figure 2.8.

Consider first the work done in lifting the weight, from 2 to 3c. This can be expressed as

$$_2W_{3c} = -\int_2^{3c} \mathbf{F} \cdot \mathbf{dx}. \tag{2.20}$$

This is the usual mechanics definition of work. Note the minus sign because work is being done on the surroundings. (The system is lifting a weight.) Notice, too, that here I have used subscripts in order to keep track of the particular process. $_2W_{3c}$ is to be read as "the work interaction as the system changes from state 2 to 3c." Because, for the piston, F is in the same direction as the displacement, x, then $\mathbf{F} \cdot \mathbf{dx} = Fdx$. Moreover $F = pA$, where A is the area of the piston, and so Equation (2.20) can be written

$$_2W_{3c} = -\int_2^{3c} pAdx \tag{2.21}$$

$$= -\int_2^{3c} pd\forall \tag{2.22}$$

$$= -m\int_2^{3c} pdv, \tag{2.23}$$

where $d\forall$, the (differential) change in volume, $\forall$, is equal to Adx because A is constant. Furthermore, $\forall \equiv mv$, and because m is constant (closed system), it has been taken outside the integral sign. It is apparent that the work in lifting the weight is proportional to the area under the curve 2–3c because this is the integral of pdv. Thus the amount of work depends on the path, and a different amount of work can be achieved for the same end states. This is why we used the δ sign in Equation (2.18), because the integral of work (and heat also) depends on the path, whereas the change in energy only depends on the end states. From this it is quite clear that the relation between p and v must be known before Equation (2.23) can be integrated. For example, if p is inversely proportional to v, that is, $p = C/v$ where C is a constant, then

$$_2W_{3c} = -m\int_2^{3c} \frac{C}{v}dv = -mC \ln \frac{v_{3c}}{v_2}.$$

The essential point is that the work done is proportional to the area under the curve on a p-v plot. This type of work is called expansion work, or pdv work. Notice, just as for our ball-bowl example, work is an interaction across a boundary. Here, a weight is being lifted external to the system because of something occurring in the system itself. Notice also that the sign of the work interaction will be negative ($v_{3c} > v_2$), indicating that work is done by the system.

For our cycle the *pdv* work from 1 to 2 and from 3c to 4 is zero because $dv = 0$. However, from 4 to 1 the work is positive because work is being done on the system. The total or net work is the area inside the *p*-*v* diagram or the difference between the work done by the system (area under curve 2–3c) and the work done on the system returning it to its initial state (area under curve 4–1). In symbolic terms

$$|W_{\text{net}}| = |{}_2W_{3c}| - |{}_4W_1|. \tag{2.24}$$

The modulus signs have been used to indicate magnitudes only. Equation (2.24) could equally well have been written as

$$W_{\text{net}} = {}_2W_{3c} + {}_4W_1. \tag{2.25}$$

Here the integral (Equation (2.23)) would show that the first term is negative. Because $|{}_4W_1| < |{}_2W_{3c}|$ (the area under 4–1 is less than that under 2–3c), then W_{net} will be negative, as it must be for the system to do work on the surroundings as it undergoes a complete cycle.

We can now calculate precisely the amount of work a piston engine does, as long as we know the paths on a *p*-*v* diagram. Note that paths 1–2 and 3–4 may not, in general, be vertical. For instance, when heat is being applied going from state 1 to 2, the pressure may be large enough to lift the weight, and thus a work and heat interaction would be occurring simultaneously. We will study this case in detail later. Figure 2.9 shows some typical heat engine cycles. There are many more.

What about the heat input required to achieve the given work? After all, the efficiency of the engine is defined as the net work divided by the heat input (Equation (1.1)). Before we can determine this we must relate heat interactions to temperature changes.

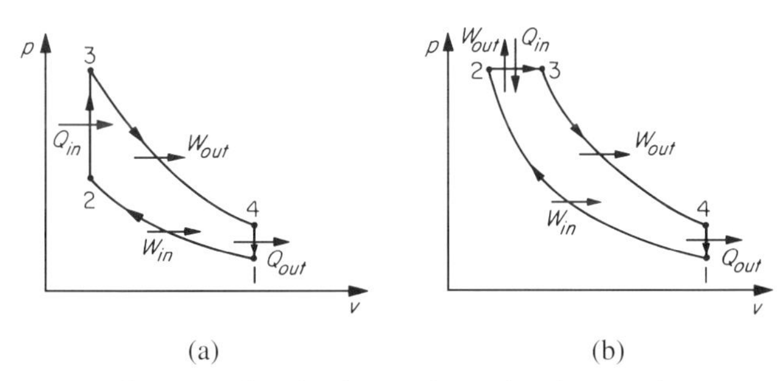

Figure 2.9 *p*-*v* diagrams for the internal combustion engine. (a) Otto cycle; (b) Diesel cycle. The Otto cycle will be discussed in more detail. The arrows indicate the direction of the heat and work interactions.

Single-phase substances, for example, air or liquid water, will change temperature when heated (providing there is no work interaction that exactly compensates for the heat interaction). However, the change in temperature for a given heat interaction depends on the nature of the substance. For instance, the heat required to raise a liter (33.8 fluid ounces) of water by 1 K is approximately three and a half thousand times that required to raise the temperature of a liter of air by 1 K. This is largely due to the density difference between water and air (1,000 kg/m^3 and 1.2 kg/m^3, respectively), but it is also partly due to the differences in molecular structure of the two substances. Determining the heat capacities and specific heats of substances, the change of T with addition of Q, posed a great problem for the classical physicists in the latter part of the nineteenth century. Its resolution came with the development of statistical physics and quantum theory. For many quantities the specific heat still cannot be determined from first principles, but it can always be measured. **Specific heat is defined as the amount of heat required to raise the temperature of a unit mass of substance by 1 K.** In general a different amount of heat is required if the process occurs at a constant volume or a constant pressure. Thus, the specific heat at a constant volume c_v is defined as

$$c_v \equiv \frac{1}{m}\frac{(\delta Q)_v}{dT} \qquad \text{or} \qquad (\delta Q)_v = mc_v\, dT, \tag{2.26}$$

where the subscript v means that the volume is held constant for this process. The units of c_v are J/(kg K). If one integrates this equation, assuming $c_v =$ constant (which is a good approximation for not too large temperature changes), then we find

$$Q = mc_v \Delta T \quad (v = \text{constant}). \tag{2.27}$$

Clearly, if c_v is large, a large heat transfer will be required to raise the temperature of a given mass by ΔT.

Similarly, the specific heat at a constant pressure is defined as

$$c_p \equiv \frac{1}{m}\frac{(\delta Q)_p}{dT} \tag{2.28}$$

Now for pdv work $\delta w = -pdv (w = -\int pdv)$ and thus the differential form of the first law (Equation (2.18)) is

$$\delta q - pdv = du. \tag{2.29}$$

If $v =$ constant then $dv = 0$, and using Equation (2.26) with $\delta q = \delta Q/m$, Equation (2.29) becomes

$$\delta q = du = c_v dT. \tag{2.30}$$

For constant c_v, Equation (2.30) may be integrated to yield

$$\Delta u = c_v \Delta T. \tag{2.31}$$

We can now determine the change in internal energy if we know the change in temperature for a constant volume process for a perfect gas. Moreover, it has been shown experimentally that for a perfect gas at moderate pressures the internal energy is only a function of temperature. From this it can be formally deduced that $du = c_v dT$ even if v is not constant, as long as it is a perfect gas at moderate to low pressures.

We are now in a position to determine the heat interactions for the cycle 1–2–3c–4 shown in Figure 2.8. For the heat addition, path 1–2, we can write the first law as

$$\delta q = du = c_v dT. \tag{2.32}$$

Here there is no work interaction because $pdv = 0$. If we assume c_v is constant, integration of Equation (2.32) yields

$${}_1Q_2 = mc_v(T_2 - T_1), \tag{2.33}$$

where $Q = mq$. T_2 and T_1 can be determined if p_2, p_1, and v_1 are known, using the perfect gas law. Thus, ${}_1Q_2$ can be determined.

In order to determine the net heat input we must go around the full cycle; thus, we next consider path 2 to 3c. Here pdv work is being done, so the first law is

$$\delta q - pdv = du = c_v dT \qquad \text{or} \qquad {}_2Q_{3c} = mc_v \int_2^{3c} dT + m \int_2^{3c} pdv. \tag{2.34}$$

Notice here that the heat addition can cause a temperature rise or do work on the surroundings, or a combination of both. Again, we require more information about the path before Equation (2.34) can be integrated. For example, assume that $dT = 0$. Here heat is being added **isothermally** (i.e., the gas inside is at a constant temperature) such that all the heat is being directly converted to work. Equation (2.34) becomes

$${}_2Q_{3c} = m \int_2^{3c} pdv.$$

Using the perfect gas law (Equation (2.19)) and recalling that T is constant, this may be integrated to yield

$$\begin{aligned} {}_2Q_{3c} &= mRT_2 \int_2^{3c} \frac{dv}{v} \\ &= mRT_2 \ln\left(\frac{v_{3c}}{v_2}\right). \end{aligned} \tag{2.35}$$

Another possibility is that the heat could have been added at a constant pressure until $3b$ was achieved, and then constant volume heat rejection could have occurred, bringing the system to state $3c$. You can determine the heat addition for this case, again using Equation (2.34).

The heat rejected going from state $3c$ to state 4 or from state 4 to state 1 can also be determined in precisely the same way as the heat input as long as the nature of the path is known. In order to determine the net heat input, Q is determined for each of the four paths and is then summed algebraically. It must be emphasized that heat and work interactions depend on the path. For the same end states, their values can be quite different. This is quite clear from the p-v diagrams for work.

You can now determine the net work and heat input for any perfect gas cycle, and thus the cycle efficiency, as long as you know the path. We will do this for some specific cycles in the next section. But how, you must be wondering, do we vary the different paths to engineer the required cycle? The answer to this problem is complex. Early on, engineers built heat engines using hit-and-miss methods and determined the path on a p-v diagram by measuring pressures and volumes. They then modified their design by, for example, varying the load on the piston as it went through its expansion, thereby changing various pressure and volume ratios. As time went on modifications led to greater efficiency. Figure 2.9 shows two of the most common cycles, the Otto cycle, which describes the automobile internal combustion engine, and the Diesel cycle, which is typical for large vehicles such as trucks and buses. Both automobile and truck engines are internal combustion engines in which a mixture of air and fuel is ignited inside the engine. During every cycle more air and fuel are brought into the cylinder and the combustion products are exhausted by means of valves (Figure 1.4). However, for these engines the ratio of air to fuel is around 15:1, and thus over 93% of the mixture is air, which can be reasonably described by the ideal gas law (Equation (2.19)). Thus, representing an internal combustion engine as an ideal gas engine with hot and cold plates instead of spark plugs and valves is a reasonable approximation and makes the analysis much easier. Notice that the hot and cold plates allow us to use the closed-system approach we have developed, whereas the real engine is an open system, with mass crossing its boundaries every cycle. We will have an example using the closed-system approach after we have considered some further aspects of cycles.

2.8 Reversibility

Given a sketch of a heat engine cycle on a p-v diagram we can now, in principle, determine the net work done from the area of the curve. We can also determine the heat interactions using the first law. From this we can calculate the efficiency of the engine. We recall that there are billions of engines in

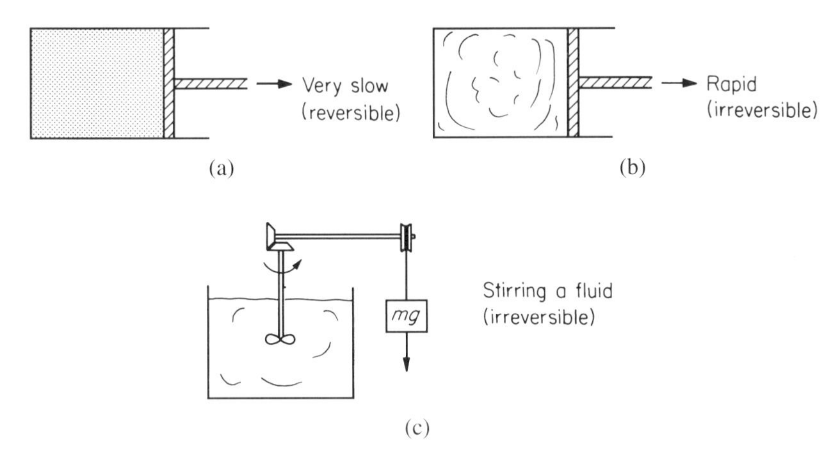

Figure 2.10 (a) Reversible expansion of a gas inside a cylinder. The process occurs slowly so that the system and its surroundings are in equilibrium at all times. (b) Irreversible expansion of a gas inside a cylinder. This process cannot be undone. (c) Another irreversible process.

the world and by increasing their efficiency by only a few percent, enormous amounts of fuel and money would be saved.

The questions we now ask are: What is the highest theoretical efficiency a heat engine can have? What will the p-v diagram look like for such an engine? Clearly, the engine can have no friction because friction will require an increase in heat input with no increase in work output; the extra heat will have to be used to overcome the friction. Because the only interactions a heat engine can have with the surroundings are work and heat interactions, in order to determine the highest possible engine efficiency, we will idealize the way these interactions take place.

Work interactions, ideally, must be frictionless. Of course it is impossible to realize frictionless motion, but we can come very close to it in the following way: If we pull a piston a small amount to the right (Figure 2.10a), and do it very slowly so that we do not cause any large-scale motion of the gas (such fluid motion, we will show in the next chapter has significant friction losses), the process may be able to be reversed, that is, it may be pushed back to its original condition so that everything returns to the same state that it started from. Such a work transfer is the most ideal we can hope to achieve. It is called a *reversible work interaction* because all properties are restored to their initial values after the pull–push motion has been completed. Of course, there must also be no friction between the cylinder and the piston, because if there were, this too would make the walls and contents of the cylinder higher in temperature, after the back and forth motion. In real life all processes are irreversible to some degree. There is some aspect of the motion that dissipates energy, so that the system's capacity for doing work is not fully utilized. Nevertheless, as we will see, the concept of reversibility plays a central role in thermodynamic analysis. Reversibility implies that everything is very close (infinitesimally close) to equilibrium at all times throughout the process. If a process is irreversible (think of very rapidly pulling out the piston so that the air at one end does not know what is happening at the other end, Figure 2.10b), there is a period of disequilibrium where the air motion is chaotic and turbulent. This cannot be undone by returning the piston to its original position. For this situation the gas inside the cylinder will be at a higher temperature after the pull–push actions, indicating the work interaction has not been properly reversed. Moreover, more work will be expended compared to the reversible case because part of the work will be used in producing the turbulent motion inside the cylinder.

The concept of reversible and irreversible work transfers does not apply only to pistons in cylinders. Consider the stirring work that occurs in a mixer (Figure 2.10c). Here there is a work interaction across the boundary because a weight is being moved (lowered) in the surroundings and thereby doing positive work on the system. (The fluid in the container is assumed here to be the system.) However, such a work interaction cannot be reversed; we cannot 'unstir' the fluid in the system by reversing the work interaction. This is clearly an irreversible work transfer. There are many others. A completely general way of determining whether a process is reversible or not will be described in Section 2.10.

Is there a heat interaction that is analogous to this reversible work interaction? Can we find a heat interaction whose direction we can reverse by a small amount, bringing both the system and its surroundings to the same condition from which they started? If the heat interaction occurs across a finite temperature difference, this is impossible. We will show in Section 2.10 that if it were reversible we could have perpetual motion. However if the heat interaction occurs at an essentially constant temperature, with only an infinitesimal difference to provide the heat transfer in the required direction, the heat transfer is reversible. If we heat the object a little and then cool it a little, the heat will flow first in one direction and then in the other in a reversible way (Figure 2.11).

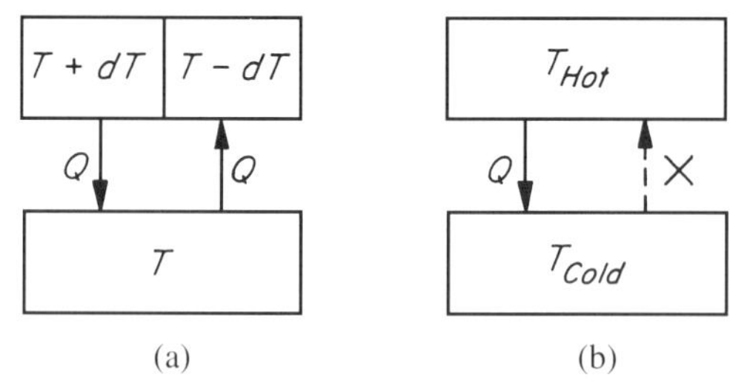

Figure 2.11 Reversible (a) and irreversible (b) heat transfer. In Section 2.10 we will show that if (b) were in fact reversible we could have perpetual motion machines.

The notion of a reversible heat transfer is just as hypothetical as that of a reversible work transfer; it can only be aimed for, never quite achieved. But you should convince yourself that heat transfer across a finite temperature difference is definitely irreversible. For this condition we cannot reverse the heat transfer by changing things an infinitesimal amount in the way we could with the reversible work transfer in the piston (a small tap in one direction or the other reversed the sign of the work interaction). For the heat transfer to become reversible we use the notion of a *limit*. We decrease the temperature difference until in the limit it is almost zero (i.e., we reduce it to an infinitesimally small difference). There still will be a heat transfer from the higher to lower temperatures. But now the direction of the heat transfer can be reversed by an infinitesimal change in the temperature step. Note that this situation departs from thermal equilibrium by only an infinitesimal amount, whereas for a finite temperature step, there is a strong thermal disequilibrium.

In summary, the concept of reversibility means that the system and its surroundings are always infinitesimally close to equilibrium. In such cases work and heat interactions can be reversed such that both the system and the surroundings will return to exactly the same states they were in before the back and forth interactions took place.

We now have a definition of reversible heat and work transfers. Now I will describe a completely reversible cycle and then show that this engine must have the highest theoretical efficiency of any engine.

2.9 The Carnot Cycle

A cycle that is completely reversible must have heat transfer occurring at (almost) a constant temperature. This implies that it must receive all its heat at temperature T_H and reject its heat at temperature T_C. The way this engine is constructed is as follows (Figure 2.12). First a hot plate, T_H, is brought into contact with a cylinder in which the gas is also at T_H. We will define the closed system as the gas inside the cylinder and will assume for simplicity that it is an ideal gas. The piston is initially far to the left, and the specific volume is v_1. A heat transfer is allowed to take place isothermally, such that the piston moves out (Figure 2.12(a)) as the heat flows in and the temperature inside the piston remains constant. (We know, of course, that the hot plate must in fact be dT greater than T_H for the heat interaction to take place. We are assuming that this difference is so close to zero that it can be ignored.) The load on the piston is carefully adjusted so that the heat and work transfers balance (so that $\Delta u = c_v \Delta T = 0$). The first law for this part of the cycle is thus

$$\delta q + \delta w = 0 \tag{2.36}$$

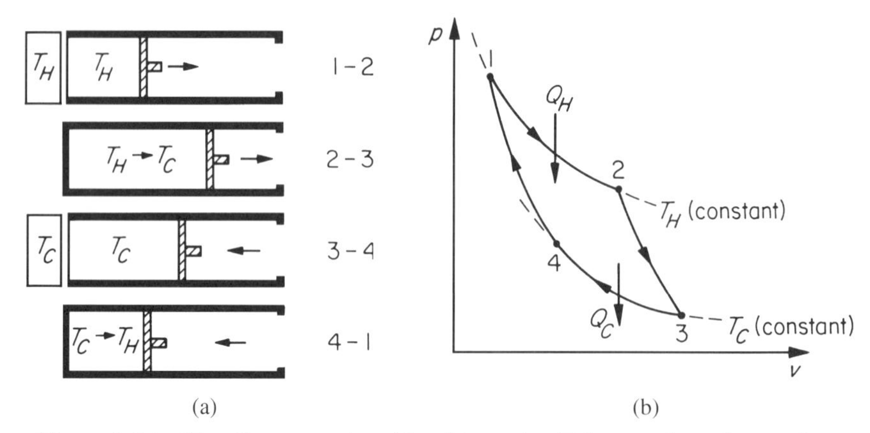

Figure 2.12 The Carnot engine. For this cycle all heat and work transfers are reversible.

or

$$\delta q = p dv. \tag{2.37}$$

Thus, using the perfect gas law,

$$Q_H = m \int_1^2 p dv = mRT_H \ln \left(\frac{v_2}{v_1} \right). \tag{2.38}$$

This type of process has been described in Equation (2.35). Notice here that both the heat and work transfers can be made reversible because, if done slowly and without any friction, pdv work is reversible also. The path of this process is shown in Figure 2.12(b). Because T = constant, pv = constant (we are assuming a perfect gas). Thus $p \propto 1/v$; that is, the path is a hyperbola. When the piston is at 2, the hot plate is taken away and the piston is allowed to slowly move out further (to v_3) with no heat transfer, an **adiabatic** expansion. Again the work transfer is reversible. The first law for the process is

$$\delta w = -pdv = du = c_v \, dT. \tag{2.39}$$

Here, T will not be constant and the total work done will be

$$W = mc_v(T_C - T_H), \tag{2.40}$$

where T_C is the final temperature of the gas in the cylinder. In order to find the shape of the path on the p-v diagram, we use Equation (2.39) and the perfect gas law. Thus

$$-pdv = -(RT/v)\, dv = c_v dT. \tag{2.41}$$

Integrating,

$$-\int_2^3 \frac{dv}{v} = \frac{c_v}{R} \int_2^3 \frac{dT}{T}. \tag{2.42}$$

Thus,

$$\left(\frac{T_3}{T_2}\right)^{c_v/R} = \left(\frac{T_C}{T_H}\right)^{c_v/R} = \frac{v_2}{v_3}. \tag{2.43}$$

Using the perfect gas law again, this can be written as

$$\frac{p_3}{p_2} = \left(\frac{v_2}{v_3}\right)^{(R/c_v)+1}. \tag{2.44}$$

For air, $R = 2.87 \times 10^2$ J/(kg K) and $c_v = 0.718 \times 10^3$ J/(kg K). For this part of the cycle the pressure decreases more rapidly than for the isothermal part. This is so because $(R/c_v) + 1 = 1.4$ (and it is greater than 1 for any gas) that is, for this process $pv^{1.4}$ = constant, whereas for the isothermal process (1–2, Figure 2.12(b)) pv = constant.

The gas has now expanded to its largest value of v and has done work for both processes 1–2 and 2–3. To complete the cycle, we reverse all processes, compressing it isothermally, with a heat transfer from the system to a cold plate at temperature T_C, and then further compress the gas, this time adiabatically, until it reaches its final state. For the isothermal heat extraction

$$Q_C = mRT_C \ln\frac{v_4}{v_3} \quad \text{or} \quad -Q_C = mRT_C \ln\frac{v_3}{v_4}, \tag{2.45}$$

and for the adiabatic compression

$$\left(\frac{T_H}{T_C}\right)^{c_v/R} = \frac{v_4}{v_1} \quad \text{or} \quad \frac{p_1}{p_4} = \left(\frac{v_4}{v_1}\right)^{R/c_v+1}. \tag{2.46}$$

The full cycle is shown in Figure 2.12(b).

Because we have found from Equation (2.43) that $(T_C/T_H)^{c_v/R} = v_2/v_3$, it follows from the first of Equations (2.46) that $v_1/v_4 = v_2/v_3$ or $v_2/v_1 = v_3/v_4$. Thus, dividing Equation (2.45) by Equation (2.38) there results

$$\frac{|Q_C|}{|Q_H|} = \frac{T_C}{T_H}. \tag{2.47}$$

Here we are interested in absolute quantities only, hence the use of modulus signs (the direction for the heat transfer is implied by the subscripts). This is a remarkable relationship because it shows that the ratio of the heat transfers is independent of the value of R and c_v (i.e., the specific nature of the perfect gas). In fact, more advanced analysis shows that Equation (2.47) holds for any reversible heat engine operating between a single high-temperature source and a single low-temperature sink, be it made with gas, water and steam, or any other substance. It is completely general.

The efficiency of the cycle is

$$\eta_{\text{th}} = \frac{W_{\text{net}}}{Q_H}. \tag{2.48}$$

Now the first law for the cycle is

$$\sum W + \sum Q = 0 \tag{2.49}$$

or

$$\sum W \equiv W_{\text{net}} = -\sum Q = -(|Q_H| - |Q_C|). \tag{2.50}$$

Note that because $|Q_H| > |Q_C|$, the net work will be negative. In accordance with our sign convention, this means that net work is done on the surroundings, as it must be. Because when we talk about an engine it is implied that work is done on the surroundings, we will use absolute quantities when referring to its efficiency, defining it as

$$\eta_{\text{th}} = \frac{|W_{\text{net}}|}{Q_H}. \tag{2.51}$$

From the first law, this becomes

$$\eta_{\text{th}} = \frac{|Q_H| - |Q_C|}{Q_H}. \tag{2.52}$$

Now, because for the Carnot cycle $|Q_C|/|Q_H| = T_C/T_H$ (Equation (2.47)), it follows that

$$\eta_{\text{th,Carnot}} = 1 - \frac{T_C}{T_H}. \tag{2.53}$$

Thus the efficiency of the reversible engine is determined solely by the temperature of the hot and cold temperature sources. On the other hand, for cycles that have irreversibilities (such as heat transfer across a finite temperature difference), the efficiency will be determined by the nature of the stuff inside the engine. Thus for the case of a perfect gas the efficiency will be a function of R and c_v (as well as temperature and pressure ratios) as you will find by doing problems 2.9 and 2.11. Note that while Equation (2.52) applies for any heat engine, Equation (2.53) only applies for the fully reversible cycle. Only for this case is the efficiency not a function of the nature of the substance that makes up the engine.

For the Carnot cycle, Equation (2.53) shows that for a thermal efficiency of unity, $T_C = 0$ or $T_H = \infty$, that is the cold plate must be at zero degrees Kelvin (-273.14 K) or the hot plate must be at an infinitely high temperature. Both are impossible to obtain; the former because heat has to be discharged into the low-temperature sink. Thus, unless the sink were infinitely large (an impossibility), it cannot be precisely at zero degrees Kelvin. So even a fully reversible Carnot engine cannot have an efficiency of 1.

It must be emphasized that our engine has no friction, it is reversible and therefore an idealization, but even for this machine, 100 percent efficiency is theoretically impossible. In the next section I will show that the Carnot efficiency is the very highest that can be achieved; all others must have an efficiency less than it. Before we do this, we will calculate the efficiencies of some reversible engines. Consider, for example, that the temperature of the hot plate is enormously high, say 8,000 K, which is the temperature of the surface of the sun. Assume that the cold plate is at room temperature (300 K). From Equation (2.53), you can show $\eta_{\text{th}} = 96.3$ percent. This is high, but not unity. More typically, T_H will be perhaps as high as 1,000°F (800 K), giving a Carnot efficiency of 63 percent. Notice that it is theoretically possible to have a Carnot engine operating across a very small temperature difference, and do work. For example on a winter day the inside temperature could be 20°C and the outside may be 0°C. We could use the warm air inside as the hot plate, T_H, to expand the piston and then take it to the window and use the cold air as the cold plate, T_C. The Carnot efficiency for this cycle would be 6.83 percent. Probably the real effects of friction would make this engine worthless.

2.9.1 A Comparison between a Real Cycle and the Idealized Carnot Cycle: An Example of How to Determine Cycle Efficiency

Let us compare a real engine, the ubiquitous internal combustion gasoline engine, to the Carnot cycle. This will also provide us with a good example of a first-law analysis of an engine. The IC engine, or Otto cycle, is described well by the p-v diagram shown in Figure 2.9(a), redrawn again in Figure 2.13(a). The cycle consists of adiabatic (no heat transfer) compression (1–2), heat addition at a constant volume (2–3), adiabatic expansion (3–4), and then heat rejection at a constant volume (4–1). As discussed at the end of Section 2.7, we pretend the heat addition

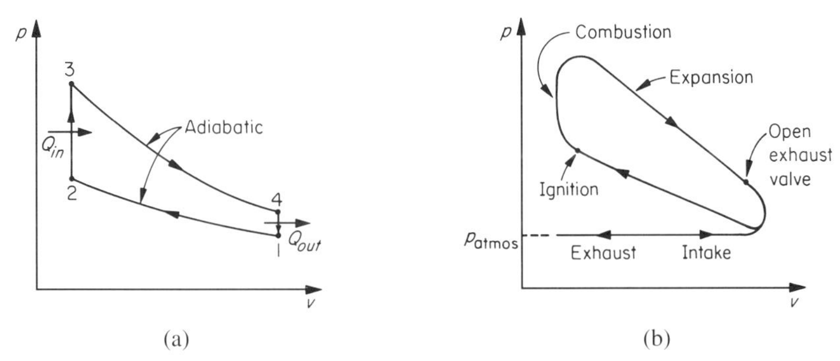

Figure 2.13 The Otto cycle. (a) The idealization of the closed-system heat engine. (b) The four-stroke cycle you would measure for a real internal combustion gasoline engine.

is by means of a hot plate suddenly applied, and the heat rejection is due to a cold plate, rather than by internal combustion (heat addition) and exhaust (heat rejection) that occurs in a real engine. If you were to measure the p-v diagram for an actual Otto engine, it would not have the nice sharp corners of Figure 2.13(a); it would look something like 2.13(b). Engineers must idealize and approximate in order to progress. Otherwise they would be lost in a miasma of empirical facts!

Assume a realistic engine with a pressure ratio, p_2/p_1, of 8:1 and that p_1 and T_1 are atmospheric pressure (1 bar $= 10^5$ N/m^2) and temperature (20°C = 293 K). The cylinder volume at the beginning of the compression stroke is 2 liter (2×10^{-3} m^3), and 4 kJ of heat are added during the constant volume heating process. We wish to determine the thermal efficiency of the cycle. R and c_v for air are 2.87×10^2 J/(kg K) and 0.718×10^3 J/(kg K) respectively.

In order to determine the efficiency it is necessary to determine the net work done (the area of the p-v diagram). This, in turn, requires knowledge of the pressure and volume at the various positions of the cycle. From the ideal gas law, we find $v_1 = 0.84$ m^3/kg. What about v_2? Because we know the path from 1 to 2 (it is an adiabatic compression), we can use Equation (2.46), which becomes

$$\left(\frac{p_2}{p_1}\right) = \left(\frac{v_1}{v_2}\right)^{(R/c_v+1)}.$$

Thus, because $p_2/p_1 = 8$, $v_2 = 0.19$ m^3/kg. From the perfect gas law, with $p_2 = 8 \times 10^5$ N/m^2 we find $T_2 = 530$ K. As expected the gas becomes hotter when compressed. Now we apply a hot plate, which raises the temperature even further, while the volume remains constant. (You can imagine that in the real engine the combustion takes place so quickly, the piston does not have time to move out.) For this process, the first law is

$$Q_{23} = mc_v(T_3 - T_2).$$

Because Q_{23} is given as 4 kJ (this would in practice be the "heating value" of the fuel) and T_2 and c_v are known, we find $T_3 = 2{,}871$ K. The mass m (which is constant) was calculated from the perfect gas law, for the conditions at 1, to be 2.38×10^{-3} kg. The large rise in temperature causes a rise in pressure. From the perfect gas law we find $p_3 = 4.34 \times 10^6$ Pa. The pressure p_4 after the adiabatic expansion is again determined from Equation 2.46. Its value is 5.42×10^5 Pa, that is, 5.42 times the atmospheric pressure. The temperature T_4 is 1,586 K.

We can now determine the net work of the cycle:

$$\begin{aligned} W_{\text{net}} &= W_{34} + W_{12} \\ &= mc_v(T_4 - T_3) + mc_v(T_2 - T_1) \\ &= -2.19 \times 10^3 + 0.405 \times 10^3 \\ &= -1.79\,\text{kJ}. \end{aligned}$$

Note that the compression work is positive (on the system) and the expansion work is negative (by the system). The net work is negative because the engine is doing work on the surroundings. The engine efficiency is then

$$\eta_{\text{th}} = \frac{1.79 \times 10^3}{4 \times 10^3} = 44.8\%.$$

The Carnot efficiency for a cycle working between the highest temperature ($T_3 = 2{,}871$ K) and the lowest temperature ($T_1 = 293$ K) of this cycle is

$$\eta_{\text{th,Carnot}} = 1 - \frac{293}{2{,}871} = 89.8\%.$$

The Otto engine does not compare well. In fact real IC engines have efficiencies even lower than we have calculated for our Otto cycle. This is due to departures from the idealizations shown in Figure 2.13(a).

2.10 The Second Law of Thermodynamics

Instead of sketching pistons, cylinders, and p-v diagrams, we will represent an engine in a more abstract way. The essence of a heat engine is that it has a heat intake from a high-temperature source (hot plate), it has a heat rejection to a low-temperature sink (cold plate), and it does net work on the surroundings. All of these characteristics are shown in the block diagram of Figure 2.14(a). The hot and cold plates are more formally referred to as high and low **temperature reservoirs**. They are considered to be sufficiently large so that their temperature remains constant over the duration of the heat transfers. The diagram implies nothing about the type of cycle; it could be reversible or irreversible. It need not be a perfect gas; indeed it could be a steam and water cycle or a cycle produced by heating and cooling rubber bands stretched over wheels. As long as it has heat interactions, does work, and goes through a cycle, anything is permissible. Now, if the cycle is reversible, then all the arrows of the diagram can be reversed. This is shown in Figure 2.14(b). (If this were a perfect gas Carnot cycle, all the arrows on the p-v diagram of Figureo 2.12 would also be reversed.) When an engine is working backwards in this way, it is called a heat pump because it is being supplied with work to pump heat from a low to a high temperature. This work, supplied to the system, can now keep the hot plate hot (or, for example, supply heat to a room to keep it at a constant (warm) temperature). And it can keep the cold plate cold (or, for example, extract heat from a refrigerator box to keep it at a constant (cool) temperature). The fact that heat is being transferred from a low to a high temperature does not violate any laws of thermodynamics because we are doing work to achieve this. (On the other hand, we will soon show, as you

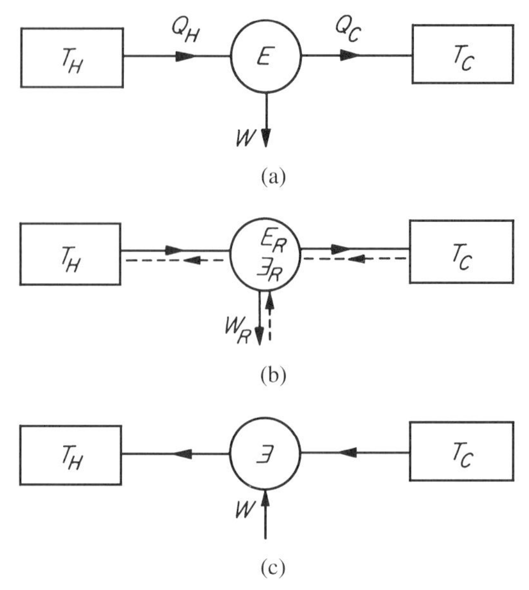

Figure 2.14 (a) Schematic diagram of a heat engine. (b) A reversible heat engine. When running in the reverse mode it is sometimes called a heat pump or a refrigerator because such an engine could keep the cold plate cold or the hot plate hot. (c) A heat pump. When an engine is working as a heat pump or refrigerator I use the symbol $\exists$. The subscript R refers to the engine being reversible. In all cases the system is the engine, with work and heat crossing its boundaries.

intuitively know, that heat, by itself cannot go from a low- to a high-temperature region.)

It is also possible to build an irreversible heat pump. This is shown in Figure 2.14(c). What is so special about the Carnot cycle is that it can go either way and when it has gone through its cycle both the system and the surroundings return to their initial states. Of course net work is done on the surroundings and there are heat interactions with the hot and cold plates, but all of these interactions are reversible. As we have already stated, most heat engines are not reversible. For example if there is friction, or if there are heat interactions over finite temperature differences, the engine will be irreversible.

We will now place a reversible engine, E_R, and another engine, E, which is irreversible, side by side, operating between the same hot and cold plates T_H and T_C, as shown in Figure 2.15(a). Both are doing work; W_R for the reversible and W for the irreversible one. Now let us size these engines so that both extract the same amount of heat Q_H from the hot plate. We will now reverse E_R, and this is done in Figure 2.15(b). We will use the symbol $\exists$ when a heat engine is working in the reverse direction. $\exists_R$ is now supplying to T_H exactly the same amount of heat that E is extracting. Thus T_H is redundant. It is merely a conducting channel, and in Figure 2.15(c) it has been discarded. We now have two engines,

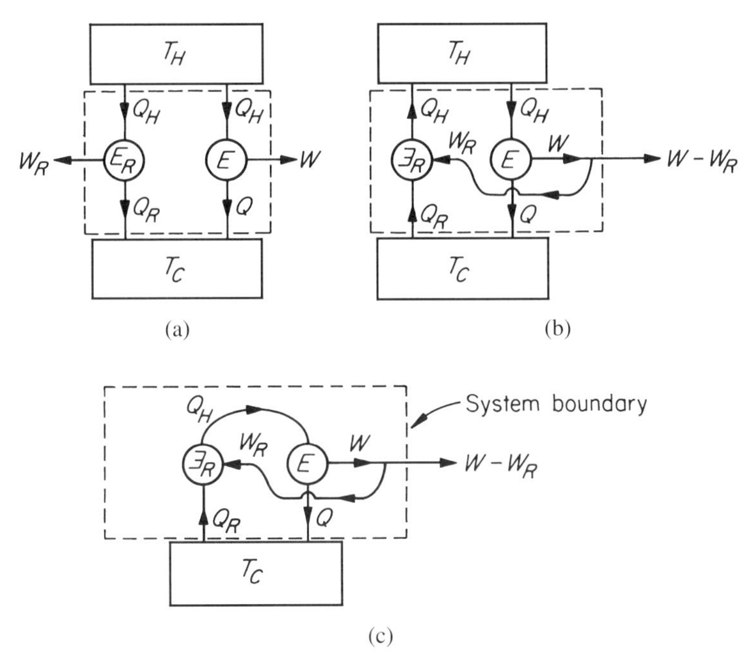

Figure 2.15 (a) A reversible heat engine (E_R) and an irreversible heat engine (E) both operating across the same temperature difference, and both extracting the same amount of heat from T_H. (b) The reversible engine (E_R) is reversed. (c) The high-temperature source, T_H, is unnecessary, and net work is produced by extracting heat from a low-temperature reservoir only.

the reversible one supplied with work and the other producing it. Let us assume for the moment that $|W|$ is greater than $|W_R|$. This means that W could be used to run $\exists_R$, and there would be some work left over ($|W| - |W_R|$) to lift a weight. The total system (Figure 2.15(c)) would be doing work on the surroundings, and it would need only a low-temperature reservoir (e.g., the surroundings). If we could build one of these systems, we could place it on our table and let it go! Because no temperature difference is required, it needs no fuel to provide a high-temperature source. Such an engine could produce perpetual motion, in violation of experience. Our energy and environmental problems would vanish. Somewhere the reasoning of our thought experiment must be wrong.

Where is the problem? Notice that the first law is not violated, because for the system (Fig. 2.15(c)),

$$-(|W| - |W_R|) + (|Q_R| - |Q|) = 0.$$

Thus, the net work out is equal to the net heat in for the cycle. We can also write the first law for each engine undergoing a cycle:

$$|Q_R| - |Q_H| + |W_R| = 0$$

and

$$-|Q| + |Q_H| - |W| = 0.$$

Because we have assumed $|W| > |W_R|$, these equations show that $|Q| < |Q_R|$. Again, this does not violate the first law.

The flaw is in assuming that the irreversible engine can produce more work than the reversible engine while operating between the same hot and cold reservoirs. If we assume $|W| < |W_R|$, everything is fine. Our system will now require work to run it. It will be a useless friction machine, converting work into heat, just like the ball-in-the-bowl example of Section 2.4, but it will not lead to an absurdity.

By means of a thought experiment we have arrived at a profound insight. Because an irreversible engine with a higher efficiency than a reversible engine would violate all we know about the world, we can state that: *It is impossible to have a heat engine with an efficiency greater than a reversible heat engine, operating between the same two temperature reservoirs.* This is known as the **second law of thermodynamics**.

Another equivalent statement of the second law of thermodynamics is the following: *It is impossible for any system to operate in a thermodynamic cycle and do net work on its surroundings while receiving heat from a single reservoir.* The system shown in Figure 2.15(c) is doing this; all the heat, $|Q_R| - |Q|$ from a single reservoir (the low-temperature source) is being converted to net work, $|W| - |W_R|$. Thus, the second law makes a clear distinction between work and heat interactions. It allows for a full conversion of work into heat, but not the reverse. The first law makes no distinction between work and heat interactions.

The second law also implies that the maximum efficiency of an engine operating between high- and low-temperature reservoirs, T_H and T_C respectively, is $1 - T_C/T_H$ (Equation (2.53)). The generality of this result must be emphasized. Although Equation (2.53) was arrived at by analyzing a perfect gas reversible cycle, the result must hold for all reversible heat engines, be they made of gas, steam, or any other substance. The result is independent of the structure of the heat engine. This follows from the thought experiment, which did not assume anything about the nature of E_R, except that it was reversible and operated between two reservoirs, T_H and T_C. So for a fully reversible engine η_{th} can only be a function of T_H and T_C. (To convince yourself that the functional dependence of η_{th} on T_H and T_C must be that of Equation (2.53), re-do the thought experiment of Figure 2.15, but now with both engines reversible. Reverse one of them and show that the second law must be violated unless both engines have the same efficiency. Now if one of the engines is a perfect gas Carnot cycle with its efficiency given by Equation (2.53), the efficiency of the other engine must be the same as Equation (2.53) no matter what it is made from.)

It is the second law, not the first, that illustrates why we have energy problems. Because all matter above zero degrees Kelvin has internal energy (because the atoms or molecules have random vibrational and other forms of energy), there is no lack of the stuff. Our silly machine of Figure 2.15 is fully consistent with the first law, converting the (internal) energy of the surroundings into work. The second law shows, however, that in order to run a heat engine we must have a temperature difference. This is why we have to have combustion and so on. Even if we took a heat engine up to the surface of the sun, we still would have to produce an even higher temperature to provide the temperature difference to run it (Figure 2.16).

Before summarizing the results of this chapter, we will prove, using the second law, that heat transfer across a finite temperature difference must be irreversible, an assertion we made in Section 2.8. We do this by assuming that the heat transfer across a finite temperature difference is indeed reversible and then show that its occurrence would lead to a violation of the second law. Figure 2.17 shows a hot reservoir, T_H, and a cold reservoir, T_C. We know that heat can go from T_H to T_C; in the sketch (on the left-hand side) we have drawn the heat going in the

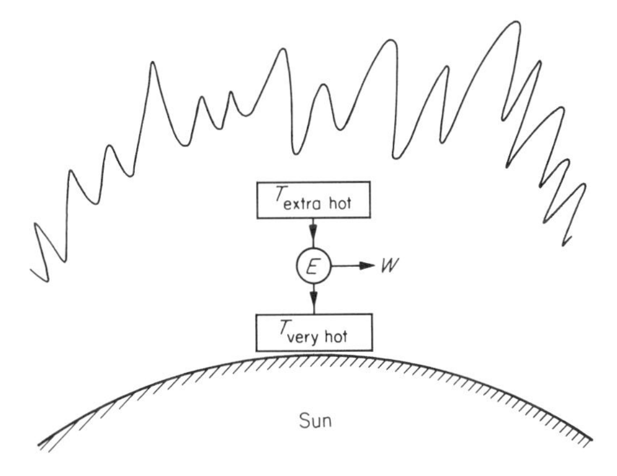

Figure 2.16 A heat engine on the surface of the sun would still require a temperature difference to run it.

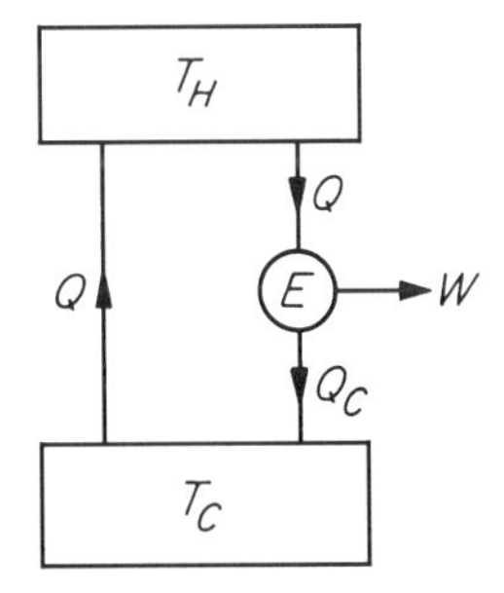

Figure 2.17 If heat could go reversibly (from T_C to T_H), this contraption shows we could have a perpetual motion machine.

reverse direction from T_C to T_H. While this suspect process is taking place, we hook up a heat engine on the right-hand side such that $W = |Q| - |Q_C|$. This heat engine is designed such that all the heat transferred from T_C to T_H is used to run the engine, which discharges Q_C to T_C. Such an engine can easily be made. Now we can eliminate T_H; it is merely a channel. Then it is clear that we have a system consisting of a heat engine (reversible or irreversible) that does net work on the surroundings and extracts heat $(|Q| - |Q_C|)$ from a single reservoir, which in this case could be the surroundings. Clearly, assuming that heat can flow reversibly between T_H and T_C leads to an absurdity – a violation of the second law and hence perpetual motion. Heat, we therefore must assert, cannot flow reversibly between finite temperature differences. This is an alternative way of stating the second law.

The reasoning we have applied to heat transfer across a finite temperature difference illustrates a general way of examining systems to determine whether they are plausible. Engineers at the patent office, for example, are often confronted with amazing and complicated new inventions, which their designers suggest will end the energy crisis. Patent engineers quickly look at the overall characteristics of the device. Is it producing more work than the net heat input? If so the machine violates the first law. Is it extracting heat from a single reservoir only? If so it is violating the second law. In either case the engine is worthless, and the generous patent engineer will return the inventor's deposit without further examination. Note that the patent engineer may have to simplify things a little before he or she will see the violation.

2.11 Summary and Discussion

The first law of thermodynamics is a budget, a book-balancing procedure. The energy of an isolated system must remain constant. If work or heat interactions occur across the boundaries, then the energy of the system will increase or decrease by the same amount as the interaction. The various statements of the first law were summarized in Sections 2.5 and 2.6.

The second law of thermodynamics places a constraint on the efficiency of heat engines. Although the first law is not violated if all the heat input to an engine is converted into work, the second law says no: Some heat must be discharged to the environment no matter how well the engine is constructed. This in turn implies that a heat engine must operate between high and low temperatures. This means we must produce high temperatures, and traditionally we have done this by burning hydrocarbon fuels. It is the second law that explains why we must burn fuel. After all, any object above zero degrees Kelvin has energy, and as far as the first law is concerned, this could be converted to work. Equation (2.53) gives an expression for the highest efficiency a heat engine can have. An efficiency of 100 percent is impossible. Note that the second law makes a fundamental distinction between 100 percent efficiency and 99.9999 percent efficiency, the latter being possible (but unlikely). Here T_H would have to be very high indeed (3×10^8 K if the low temperature was 300 K). But for 100 percent efficiency absurdities arise: Either $T_H = \infty$ K or $T_C = 0$ K.

You may be somewhat alarmed by the categorical nature of the statements I have made in this chapter. After all, engineers and scientists should be cautious. Rutherford said less than 70 years ago (in 1933), that it would be impossible to produce energy by splitting the atom, and Freud thought it would be impossible to produce a birth control pill. Yet here we are applying similarly bold statements to the upper efficiency of heat engines.

The second law of thermodynamics has deep implications. As we have shown (Equation (2.53)), Carnot and his fellow engineers determined the upper limit of a heat engine efficiency to be

$$\eta_{\text{th,Carnot}} = 1 - \frac{T_C}{T_H}.$$

This equation is remarkable on two counts. First it does not say anything about the nature of the material the engine is made from. It applies to any reversible heat engine operating between two constant temperature reservoirs, T_H and T_C. We have derived it for a special engine, a piston and cylinder with a perfect gas inside it. However the engine could have had water and steam inside it, or it could have been constructed on entirely different principles, say with rubber bands (being heated and cooled) so that they moved wheels. Equation (2.53) is independent of the properties of the material, and in more advanced courses it is derived in a completely general way. It says that if you want a heat engine that goes through a cycle and does work, it can never have a higher efficiency than unity minus the ratio of the coolest to hottest temperature in the cycle. If you drive past a massive power generation plant with steam turbines and nuclear reactors and the like, ask the chief engineer what is the highest temperature anywhere in the plant, and what is the lowest temperature. You will then, in the face of this enormous complexity, be able to give an upper bound to the overall thermal efficiency using Equation (2.53). In practice, of course, it will be much lower than your answer.

Second, the second law says something very fundamental about the nature of matter itself. This apparently contradicts the previous paragraph, where I stated that the second law does not say anything about the nature of the material the engine is made from. Let me explain. By showing that a heat engine cannot convert all the heat into work (while the opposite of course is possible), the second law indicates a strong distinction between heat and work, a distinction not made by the first law, which treats them as equals. Roll a pencil along a table top. After a moment it will stop. Friction, we say. We have done work by pushing the pencil to set it in motion; as it moves it rubs along the table and converts its kinetic energy into internal energy. The pencil, table, and surrounding air increase in temperature by a small amount, just as for the ball in the bowl. Sit and watch. Watch for a long time. Wait for the pencil to move back to where it started from or for the ball to move up the side of the bowl by itself. For it to happen, all of the atoms (or molecules) vibrating now slightly faster in the pencil, table, and surrounding air would have to decide to vibrate in unison, in one direction, instead of in their random way. If they could do this the pencil could move back to where it came from. Watch for 10^{10} years, the age of the universe. Probability theory says its chances of happening are very very low indeed. Our own experience tells us it is zero. When the second law of thermodynamics was first enunciated, the notion of matter being made up of atoms (let alone them having random vibrational and translational modes) was a very conjectural idea. But as you see, the second law leads us to consider the very nature of matter itself. The second law, then, while not saying anything about the nature of the substances that we are dealing with, implies that they are made up of atoms in random motion and that when heat is added, the random energy of the atoms increases. The random energy cannot be directly converted back into work.

It is this remarkable generality of the second law that places it on such fundamental footing. It is as fundamental as the notion of atoms and molecules themselves. This is why I have been so emphatic. In order to solve our energy crisis, clearly we must work within the constraints of the second law.

2.12 Problems

2.1 Defining a system and then carefully observing the interactions across its boundaries is one of the central techniques of thermodynamics. As an illustration, consider the following example (from Spalding and Cole, "Engineering Thermodynamics," Arnold, 1958).

Two identical glasses are each half filled, one with white wine and the other with red wine. A teaspoon of red wine is transferred from the red wine glass into the white wine glass. This is mixed and then a teaspoonful of this mixture is transferred back into the red wine glass and this is also mixed. Which of the resultant mixtures is the more pure? If you define your systems carefully, you will be able to work this problem out in a few lines without any algebra.

2.2 (a) A tank containing a fluid is stirred by a paddle wheel and is heated at the same time. The work input by the paddle wheel is 5,000 kJ. The heat input to the tank is 1,500 kJ. Considering the fluid in the tank as the system, determine the change in internal energy. (Write down the first law of thermodynamics.) What would you do to return the system to its initial state?

(b) A system changes state in a process 1–2, and a second process 2–1 returns the system to its initial state along a different path. Fill in the blanks for the two processes. (The energy unit is the same for all quantities.)

Process	Q	W	ΔE
1–2	−10	4	
2–1	3		

2.3 Redo the problem of the ball in the bowl (Section 2.4) but now with the ball as the system. Determine the work and heat interactions that occur when the ball rolls from the top of the bowl to the final position at the bottom. Just as for the system when the box was the boundary, you must assume something about how the boundary of the new system (the surface of the ball) behaves. Assume that although friction is present, the ball is essentially nonconducting (assume it is made from wood). Thus the temperature of the ball will not change, but that of the bowl will, because of friction. (Assume the bowl is made of a conducting material such as copper.)

When you have done this problem, reverse the situation, making the ball copper and the bowl wood.

Hint: If the boundary of a system is nonconducting, there can be no heat transfer across it. Note that this problem shows that the definition of work is broader in thermodynamics than in particle mechanics.

2.4 Figure 2.18 shows a 1-kg mass suspended in a box. The cord is weightless, and the pulleys are frictionless. The outside weight is infinitesimally lighter than the inside one, and it is held down by a latch. The box is the system boundary.

(a) Assume the outside weight is unlatched and slowly the inside weight descends to the bottom of the box, landing gently on the bottom. What are the values of the work and heat interactions and the change in energy of the system?

(b) Now, instead of letting the weight move slowly down to the bottom, cut the string, letting it crash to the bottom. Determine the work and heat interactions and the changes in energy of the system.

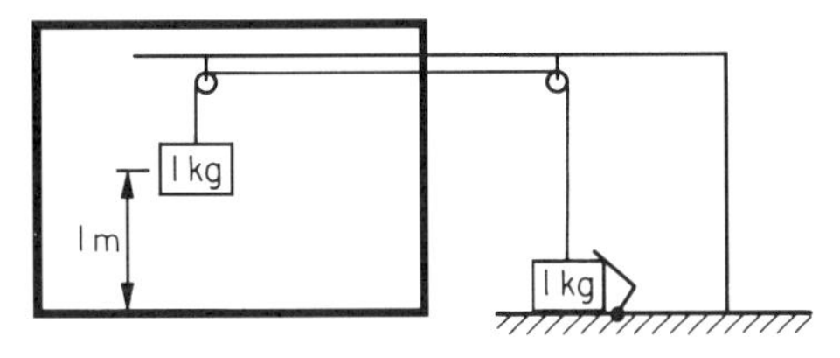

Figure 2.18 Sketch for Problem 2.4.

Hint: For part (b) we must say something about the nature of the system boundary. Assume first that it is nonconducting (therefore the system is isolated in terms of a heat transfer) and then that it is conducting. (In part (a) we need not worry about the nature of the box. Why?) Also, in doing part (b) it may help if you imagine the string to be cut from inside the system.

2.5 A frictionless piston-cylinder containing a perfect gas ($pv = RT$) undergoes a process in which the specific volume, v, doubles and the pressure, p, is halved. The process can take two different paths:
(a) T = constant
(b) Tv = constant until the final volume is attained and then v = constant.
  (i) Draw the process on a p-v diagram.
  (ii) Calculate the pdv work done for each case, and determine their ratio.
  (iii) For which process is there the largest heat interaction? (Note that the end states are the same.) What is the sign of Q? Assume that potential and kinetic energy changes are negligible.

2.6 When a fluid is stirred, although its internal energy increases because of the internal friction of the fluid motion, and thus its temperature will rise, the action of stirring is a work interaction. This is so because it complies with our definition, which is that work is an interaction between a system and its surroundings when the sole effect external to the system could be the lifting (or lowering) of a weight. Figure 2.19 shows a stirring experiment. If the specific heat of water $c = 4.2 \times 10^3$ J/(kg K) (for water, which is incompressible, $c_p = c_v$) and the container has 1 liter of water (10^{-3} m^3) of density $\rho = 1{,}000$ kg/m^3, how far would a 5-kg weight have to fall for

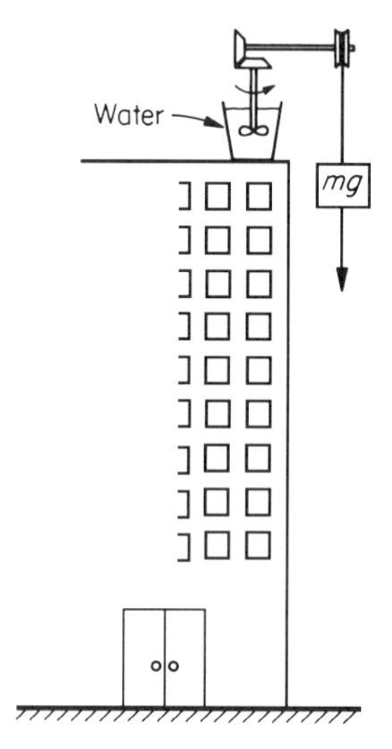

Figure 2.19 Sketch for Problem 2.6.

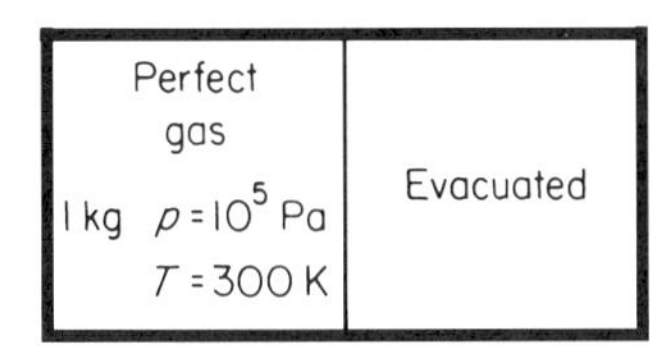

Figure 2.20 Sketch for Problem 2.7.

the water temperature to rise by 1 K? Assume the container is insulated. Can you imagine this stirring work being undone?

2.7 Consider the insulated box shown in Figure 2.20. One kilogram of a perfect gas with a pressure of 10^5 Pa and temperature of 300 K fills the left-hand side, which is separated from the evacuated right-hand side (of the same volume) by a partition. Determine the pressure and temperature of the gas after the partition has been pulled out and the gas fills the entire volume. (Begin by writing the first law for the process. Assume the work in pulling out the partition is negligible.) Is this process reversible? Can you imagine it being undone?

2.8 Consider a piston-cylinder system in which the gas expands at a constant pressure and lifts a weight. Show that the first law for this process is

$$ {}_1Q_2 = (U_2 + pV_2) - (U_1 + pV_1), \tag{2.54} $$

where 1 and 2 are the initial and final end states. The group $U + pV$ (or $u + pv$) often occurs in thermodynamics and is called the enthalpy, H (or $h \equiv H/m$). It follows from Equation (2.54) and Equation (2.28) that

$$ c_p = \left(\frac{dh}{dT} \right)_p . $$

Show that for a perfect gas

$$ c_p - c_v = R. $$

2.9 Determine the thermal efficiency of the Diesel cycle (Figure 2.9(b)), and compare it to the efficiency of an Otto cycle (Figure 2.9(a)) having the same pressure ratio (p_2/p_1) and the same volume ratio (v_2/v_1). Which has the greater efficiency? Which does the most work?

2.10 You may have seen a feeding duck like the one shown in Figure 2.21. When the beak of the duck becomes wet, the fluid inside the glass body and neck expands, changing the center of gravity and allowing the bird to bob up and down forever. Show that this is in fact a heat engine undergoing a cycle. What are the high and low temperatures of the cycle? (If you have not seen this duck before you should realize that the fluid inside has the special property of expanding as a result of a change in temperature.)

Figure 2.21 Sketch for Problem 2.10.

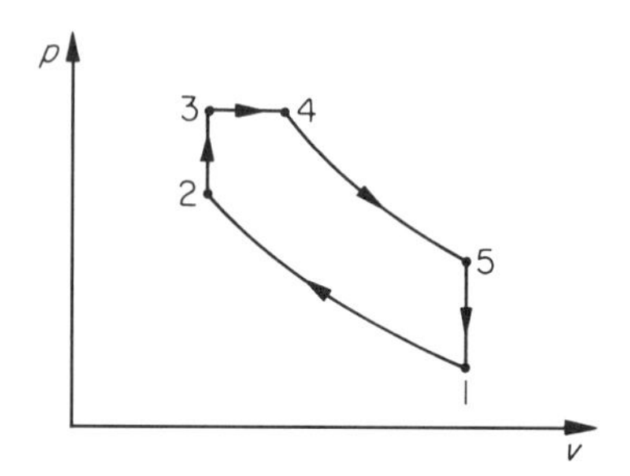

Figure 2.22 Sketch for Problem 2.11.

2.11 The *mixed* or *dual cycle* is shown in Figure 2.22. It consists of heat addition at a constant volume (2–3) followed by heat addition at a constant pressure (3–4). The rest of the cycle is the same as the Otto or Diesel cycles (Figure 2.8). Show that the thermal efficiency is

$$\eta_{\text{th}} = 1 - \frac{T_5 - T_1}{(T_3 - T_2) + \gamma(T_4 - T_3)},$$

where $\gamma = c_p/c_v$, the ratio of the specific heats.

2.12 We have shown that an engine must run between a high-temperature source and a low-temperature sink, emphasizing the importance of the temperature difference; the greater the temperature difference, the greater the efficiency. But consider a heat engine with, say, $\Delta T = T_H - T_C = 200$ K. Determine its Carnot efficiency if it were running on the earth ($T_C = 300$ K) and if it were running on the surface of Venus ($T_C = 500$ K). Can you explain why, from a physical viewpoint, these efficiencies are different?

2.13 Experience tells us that the processes occurring in Problems 2.6 and 2.7 are irreversible. Reversing the stirring will not decrease the temperature of the water (example Problem 2.6), nor will the gas spontaneously rush to the left-hand side of the box (example Problem 2.7). It can be proved that both of these processes are irreversible by employing the same method that was used to show that heat cannot flow from a low to a high temperature without any other interactions (Section 2.10). There we assumed that the heat could go from low to high temperatures and showed that if it did, the second law would be violated. Use this technique to show that the processes in Problems 2.6 and 2.7 are irreversible. Be patient, it is not easy!

The processes in these problems are known as stirring work and unrestrained expansion and are often used as examples of irreversible processes. Just as with heat transfer across a finite temperature difference, if either of these processes occurs in any part of a cycle, automatically the cycle must have an efficiency less than the Carnot efficiency.

2.14 Two heat engines, A and B, have the same efficiency and the same difference in reservoir temperatures but do not operate between the same two reservoirs. The hot-reservoir temperature of A is greater than that of B.

(a) Can both heat engines be reversible?

(b) Can both heat engines be irreversible?

(c) If only one heat engine is reversible, which one is?

Symbols

A	area	m^2
c_p	specific heat at a constant pressure	J/(kg K)
c_v	specific heat at a constant volume	J/(kg K)
E	total energy of the system	J
F, $\mathbf{F}$	force	N
g	acceleration due to gravity	m/s^2
H	enthalpy	J
h	enthalpy per unit mass	J/kg
M	molar mass or molecular weight	kg/(kmol)
m	mass	kg
p	pressure	Pa
Q	heat interaction	J
Q_{net}	heat interaction for a full cycle	J
q	heat interaction per unit mass	J/kg
R	gas constant	J/(kg K)
R_u	universal gas constant	J/(kmol K)
T	temperature	K
U	internal energy	J
u	internal energy per unit mass	J/kg
V	velocity	m/s
v	specific volume	m^3/kg
W	work interaction	J
W_{net}	work interaction for a full cycle	J
w	work interaction per unit mass	J/kg

z	height	m
$\forall$	volume	m^3
Δ	change in quantity	
δ	differential change in heat or work transfer	
η_{th}	thermal efficiency	
ρ	density	kg/m^3
$\sum$	summation	

Subscripts C and H refer to the cold and hot plates respectively.

3

Fluid Dynamics

Jet streams, billowing cumulus clouds, tornadoes, and hurricanes are some of the atmospheric motions that we are familiar with, either from direct experience or from satellite and aerial pictures. Less familiar, but equally exciting, is the motion inside an internal combustion engine. Here too there are vortex motions like mini hurricanes, jets of fluids, and convective activity. All of this takes place in the space of a few thousandths of a cubic meter (Figure 3.1). Atmospheric motion and the motion inside an engine, as well as the motion inside the sun and galaxies, the breaking of ocean waves, and the flow of blood inside our veins and arteries, are the subjects of fluid dynamics – of central importance not only to mechanical engineers but also to civil, aerospace and chemical engineers, oceanographers, cosmologists, meteorologists, and physicists. One of the most remarkable aspects of fluid dynamics is that the equations that describe the turbulent motion of a cloud or a waterfall are the same as those that describe the lubricating film in a bearing, the motion of gases in a jet engine, or the flow of blood in our veins. It is the purpose of this chapter to introduce you to some of the richness of this subject and to show you that in order to design a better engine, as well as to understand environmental pollution, we must have a deep understanding of fluid motion.

In Chapter 2 we determined the efficiency of an engine from knowledge of its cycle. In particular we determined the highest possible efficiency an engine can have, the Carnot efficiency. It is remarkable that we did this without saying anything about the nature of the fluid motion inside the engine, apart from the fact that it had to be reversible: Everything had to be close to equilibrium at all times, and thus there could be no abrupt changes in the motion. Real engines, of course, move very fast. Typically the pistons move up and down over 20 times per second. The fluid motion is turbulent, and frictional effects are important. Although thermodynamics provides the tools for determining the efficiency of

(a)

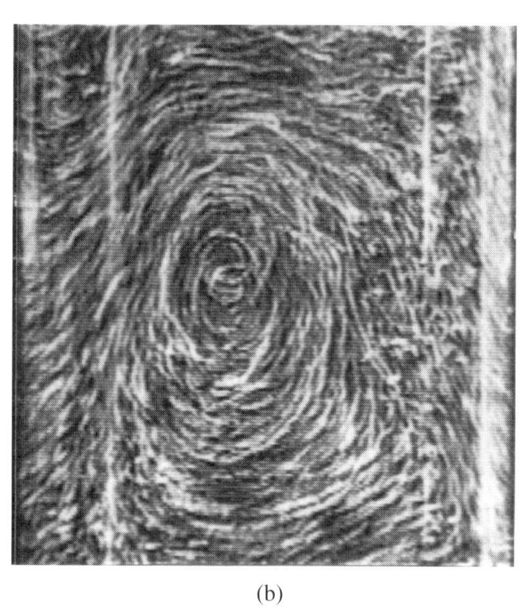

(b)

Figure 3.1 Fluid flow in the atmosphere, (a), and in an internal combustion engine, (b). The water vapor of the clouds traces the fluid motion in the atmosphere, and flow visualization techniques using small particles show it in the engine. (Part (a) courtesy of NASA. Part (b) courtesy of Dr. D. Haworth, General Motors, and reprinted with permission from SAE paper No. 941871 © 1994 Society for Automotive Engineers, Inc.)

an engine given its cycle, it does not tell us how to reduce frictional effects, or optimize the heat transfer, in order to make a given engine more efficient. This is the problem of fluid mechanics (and heat transfer and combustion, which will be dealt with in Chapters 4 and 6). Of course, thermodynamics does provide the upper bound for engine efficiency, telling the design engineer when his or her job is just about done. Thermodynamics is a little like a black-box activity. Given the inputs and outputs, the first and second laws will provide the engine efficiency without knowledge of the complex dynamics inside. In order to make better engines and better power stations and a cleaner global environment, we must understand the details inside; we must open the black box.

We will begin our study of fluid dynamics by examining the causes of fluid friction. We will show that the frictional effects of fluid motion are always present, but they increase dramatically when the motion changes from smooth, orderly laminar flow to chaotic, disorderly turbulent flow. Fluid friction causes pressure drops in pipes (and thus the need for the very large pumps in oil and gas pipe lines) as well as irreversibilities in engines and the like. Our purpose is to show why friction occurs and how to describe it.

We will then develop the Bernoulli equation in order to study the energetics of fluid flow. This will lead us to a statement of the first law of thermodynamics for an open system: a system in which mass, as well as heat and work, crosses the system boundaries. (Remember, the closed systems of Chapter 2 allowed for heat and work interactions only across the system boundaries.) Gas turbines and jet engines are open systems, and the principles of their operation will then be outlined. We will end the chapter by returning to the internal combustion engine, describing the fluid motion inside it. Thus, although I will be providing a general introduction to fluid dynamics, the focus will be on engines.

Before we begin, we must say a few words about what a fluid is and how it is defined.

There are four states of matter: solids, liquids, gases, and plasmas. The first three are quite familiar; plasmas are somewhat like gases with the electrons stripped from the nucleus of the atoms. They are the stuff of the upper atmosphere (the ionosphere) and the stars and galaxies. Fluid dynamics covers the motion of liquids, gases, and plasmas. No wonder it is an important subject; well over 99.999999 percent of the universe is in fluid motion! Here we will confine ourselves mainly to liquids and gases. We formally define a **fluid** as a substance that cannot resist a shear stress; that is, a force per unit area that is applied tangentially (rather than normally). Solids, by contrast, can resist shear stresses. If they could not your pad would always slip away from you as you write. The concept of shear stress will be studied in some detail in the pages to follow.

3.1 Smooth Streams, Transition, and Turbulence

How many times a day do we turn on a faucet? Do it now, first do it very slowly, and you will see glassy, orderly flow. If there is no wind or other disturbance, nothing will change. This is called **laminar flow**. A photo taken now will be identical to one taken half an hour later. Such a flow is deterministic; information about its future behavior is completely determined by specification of the flow at an earlier time. Now open the faucet to full on, or better still open a fire hydrant, or watch a smoke stack. Here, for this faster or larger-scale motion, the flow pattern is changing all the time. Although its **mean** motion is in one direction (sideways for the fire hydrant, up for the smoke stack), within the flow there are irregularities everywhere. For example, if you could train your eyes on a small speck of dust it would certainly move along, but it would jitter as well, sometimes darting to one side or up or down. Turbulent flow, although proceeding in a particular direction like laminar flow, has the added complexity of random velocity fluctuations. The flow patterns never repeat themselves (Figure 3.2). To convince yourself of this watch a smoke stack for a few minutes.

(a) (b)

Figure 3.2 Orderly, deterministic laminar flow and turbulent flow for which the details of the motion are never identically repeated. The laminar flow in (a) is of a gas flame. The turbulent flow in (b) is a jet of water flowing into water. (Part (a) courtesy of Professor R. J. Santoro, Pennsylvania State University. Part (b) courtesy of Professor K. R. Sreenivasan, Yale University.)

Fluid flow that is slow tends to be laminar. As it speeds up a **transition** occurs, and it crinkles up into complicated, random turbulent flow. But even slow flow coming from a large orifice can be turbulent; this is the case with smoke stacks. Engineers and scientists don't like to say "fast" and "slow" or "small" and "big" because there is no reference. Small compared to what? Big compared to what? Because turbulence is altogether a different type of fluid flow from laminar flow, it is desirable to be able to quantify under what conditions it occurs.

Let us redo the faucet experiment in a more systematic way. We have shown that as the speed, V, increases, transition to turbulence will occur. Now, instead of using water in your pipes, replace it with honey. Assuming you could provide a large enough pressure, even for fast flow the motion would remain laminar. If you do not wish to do this experiment, move a spoon rapidly in a cup of water, and then at the same speed (working hard!) in a cup of honey. Honey has a higher **viscosity** than water, and the viscosity resists transition to turbulence: while the water is turbulent, the honey remains laminar at the same speed. Finally, put a nozzle on your tap and constrict the water flow into a fine glass capillary tube. Here too the flow can be made to go quite fast without it becoming turbulent. Our experiments suggest that laminar flows occur for low speeds, small diameters, and high viscosities, whereas turbulent flows occur for the opposite conditions: high speeds, large diameters, and low viscosities. Now, viscosity is a measurable fluid property (as its density, temperature, etc.). We will discuss it more in a moment, here we will give its symbol, which is ν, and its unit, which is m^2/s. Notice that its dimensions are the same as a length multiplied by a velocity. If the fluid speed is $V(m/s)$ and the orifice diameter is $d(m)$, then we can write the following dimensionless ratio

$$Re = \frac{Vd}{\nu}. \tag{3.1}$$

Re is the Reynolds number (Figure 3.3), named after Osborne Reynolds who did systematic experiments, of a similar type to those described above, 100 years ago. Notice that if V or d (or both) is small and ν is large, *Re* will be small. For this case the flow will be laminar. Increase d or V or decrease ν, and the *Re* will increase. Reynolds found that for flow in a pipe it did not matter which of the three particular parameters he varied in this dimensionless group: As long as *Re* was less than approximately 2,000, the flow was laminar. Above this value, turbulence would invariably occur. This is a general result because it allows us to vary the type of fluid, flow speed, and pipe diameter without having to use the words *large* or *fast*, etc. Moreover, because the *Re* is dimensionless, it does not matter which system of units is used (SI, British, etc.) so long as they are the same throughout. We can now talk of high-Reynolds-number pipe flow or low-Reynolds-number pipe flow, knowing that in this context "low" means somewhat less than 2,000. The viscosity of water is approximately 10^{-6} m^2/s

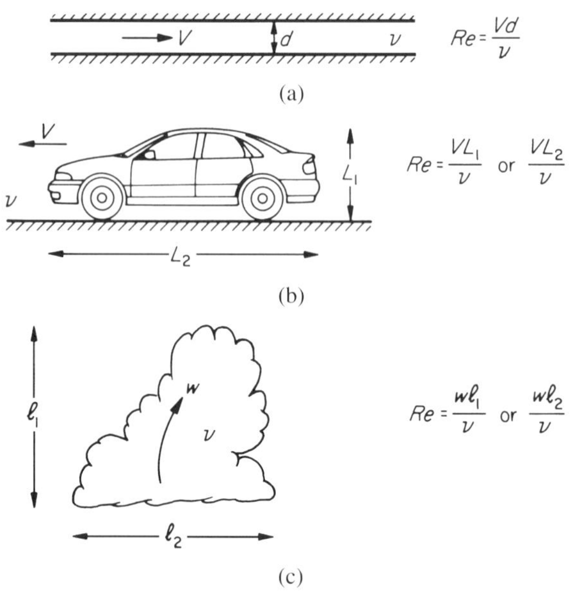

Figure 3.3 The Reynolds number, *Re*, is always defined as a velocity times a length divided by a viscosity. (a) For flow in a pipe, the length scale is the diameter (or sometimes the radius) and the velocity is the average flow speed (see Section 3.2). The length of the pipe is irrelevant because doubling it will not change the nature of the flow in the pipe (providing the pipe is long enough to begin with). (b) For the automobile there are at least two possible length scales. But note that they are the same order of magnitude, and so too will be the *Re* based on either of the two lengths. Whether the *Re* is 2×10^6 or 4×10^6 does not matter. (c) Here we have defined the *Re* in a cumulous cloud based on the convective velocity inside the cloud. As for (b) there are two length scales, but they are again of the same order. For the flow over the automobile or in the cloud the *Re* is very high, so the motion will be turbulent. The flow in the pipe could be laminar or turbulent. Osborne Reynolds showed that if the *Re* is less than 2,000 the flow in the pipe will be laminar; above that the flow tends to become turbulent.

(that of honey is about 10^{-3} m^2/s, 1,000 times greater than that of water). Thus if the pipe diameter is, say, 1 cm, the speed at which the Reynolds number is 2,000 is, from Equation (3.1), 0.2 m/s or approximately 0.4 mph, a rather slow speed. Water undergoes transition to turbulence at low speeds. Most of the water flows we see, such as in streams and rivers, are indeed turbulent.

Air too is a fluid, and its viscosity, ν, is approximately 10^{-5} m^2/s. This is a higher viscosity than that of water. This rather unintuitive fact is due to the great differences in density between the two fluids. Water has a density of approximately 1,000 kg/m^3; the air density is 1.2 kg/m^3. Thus, part of the viscous feeling we have when we pull our fingers through water is really due to inertia – we are having to move the water away from our hands, and this also provides resistance. But if we separate out this inertia effect, the viscous resistance due to the relative motion of the water molecules is in fact less than that due to the relative motion of the air molecules. This is reflected in the lower value of ν for water than for air. If we wish to include inertia we multiply the viscosity of a substance by its density; this is called the dynamic viscosity $\mu \equiv \rho\nu$. The dynamic viscosity of water is approximately 10^{-3} kg/(m s) whereas that of air is 1.7×10^{-5} kg/(m s). Thus, the dynamic viscosity of water is higher than that of air, in keeping with our intuitive notion. ν is known as the kinematic viscosity; it can be thought of as a normalized dynamic viscosity.

Although the transition from laminar to turbulent flow occurs at a Reynolds number of approximately 2,000 in a pipe, the precise value depends on whether any small disturbances are present. If the experiment is very carefully arranged so that the pipe is very smooth and there are no disturbances to the velocity and so on, higher values of *Re* can be obtained with the flow still in a laminar state. However, if the *Re* is less than 2,000, the flow will be laminar even if it is disturbed. Thus 2,000 is the value for *Re* below which turbulence will not occur in a pipe. Moreover, if the flow has a different geometry, such as flow in a square duct or over a turbine blade, transition will occur at different values of Re. The essential point is that flows become turbulent at high Reynolds numbers, where "high" means much greater than unity.

Air motion is invariably turbulent. Consider a smokestack (which to a first approximation is mostly air). If its diameter is say 3 m, then V must be less than 6.6 mm/s (0.015 mph) for it to be laminar! (We have taken ν of the smoke to be 10^{-5} m^2/s, the same as air.) There is no such thing as a laminar smokestack. Clouds too are usually turbulent. Here we determine the Reynolds number using an approximate characteristic dimension of the cloud such as its height or width (both are generally of the same order; Figure 3.3). Assuming the cloud dimension is, say, 500 m and its characteristic internal motion is, say, 5 m/s, then taking $\nu = 10^{-5}$ m^2/s (it is approximately the same for water vapor as it is for air), the $Re = (500 \times 5)/10^{-5} = 2.5 \times 10^8$, a high value indeed. No wonder cumulus clouds always have a random, puffy-looking turbulent structure. Note that I did not use the mean motion of the cloud as a whole to determine the Re. This has nothing much to do with what is happening inside the cloud. I will discuss more about characteristic velocities in Section 3.6.

3.1.1 *Why Is It So?*

Why does fluid flow manifest itself in such a strange way? Why does orderly flow turn into random turbulent flow at high Reynolds numbers? Although there have been volumes of research on this problem over the past century, many aspects of this problem still remain unsolved. We do understand the broad principles and have developed theories that predict the flow state at which the transition just begins. But predicting the details of the transition process itself is difficult and has been achieved only for a few types of fluid flows. There are good reasons for this. The transition process is chaotic. Intuition should tell us that such problems are difficult. Sociologists, political theorists, and economists also find that prediction of transition from one clearly defined state to another is difficult. It is easier to predict the next day of the stock market, or emigration patterns, if a country is in a stable state with everything going on from day to day roughly as before, than it is to predict the next day if the country is at the brink of revolution or war. Here political theorists and the like are capable of explaining what has happened only after they know the result! Fluid flow transitions are a little like this. Indeed, the transition from laminar flow to turbulence is quite like going from a gentle peace to the chaos of war! However, in contrast to political science, in fluid dynamics we know the governing equations. These are Newton's laws of motion applied to every position in the fluid. When the fluid undergoes transition to turbulence, each part of the fluid moves at different speeds and directions relative to its neighbors. Thus, although in principle the details of the motion of all parts of the fluid could be determined from one moment to the next, in practice this becomes impossible because there are billions and billions of equations to solve simultaneously as the various parts of the fluid change relative to their neighbors. In order to deal with this, methods from statistical mechanics must be used. Before we can further address the problem of transition and turbulence, we must learn a little more about fluid mechanics.

3.2 Why We Need Pumps: Friction and Stresses in Fluids

3.2.1 Laminar Flow

Instead of dealing with water coming out of the faucet, let us examine its motion in a long stretch of smooth pipe. First consider laminar flow. If the pipe is made of clear plastic or glass, we can put specks of dye close to the wall and at various distances across the pipe. Now, if we do this when the fluid is moving at a constant flow rate, we notice a remarkable thing. The dye close to the pipe wall does not move, whereas that in the center moves the fastest. A very small speck of dust would move along fastest in the center and hardly at all close to the wall. Detailed measurements of laminar flow show that the variation of velocity across the pipe is a parabola, with zero velocity at the pipe wall (Figure 3.4(a)). It is a remarkable fact that most fluids, such as air, water, oil, and blood, do not slip along solid surfaces. Our own observations tend to confirm this. Water can flow very fast in a stream, yet you will find moss on the pebbles and stones. If the velocity did not slow near the stones the moss would be pushed off. Or try to blow crumbs from a bread board. There will always remain a thin film of crumbs at the surface. The complete slowing down of a fluid along a solid surface is

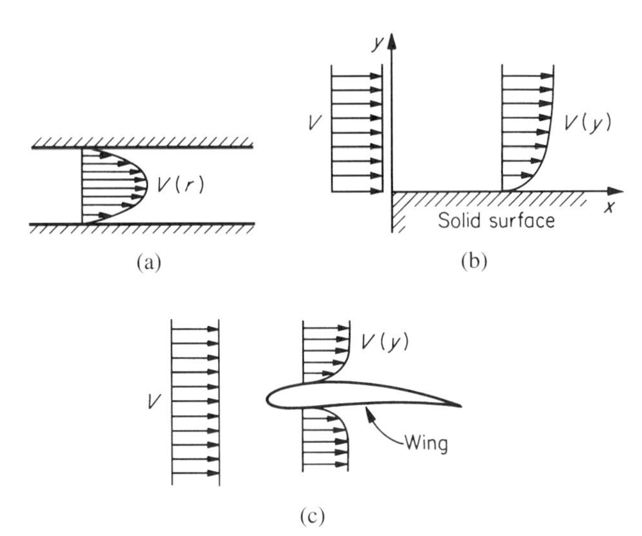

Figure 3.4 (a) The velocity profile for laminar flow in a pipe. At the wall the velocity is zero. The actual form of the profile is a parabola. (b) and (c) The velocity profile for flows over other objects. Be the flow laminar or turbulent, the velocity is zero (for most fluids) at the solid surface, although the shape of the velocity profile will depend on the Reynolds number and geometry of the object.

known as the **no-slip condition**. It is a fundamental characteristic of fluid flow. The physics of why it occurs is complex. At the molecular level, the smooth wall is really rough, and the fluid molecules hitting the surface are absorbed by the surface for a moment. As they emerge they have lost their original sense of direction (Figure 3.5). In Figure 3.4(b) and (c) we show some velocity profiles for flows other than in a pipe. No mater what the flow, the no-slip condition holds at the wall, for most fluids. There are rare exceptions, such as helium at temperatures very close to 0 K.

The no-slip condition forces a velocity gradient on the pipe flow, that is, a change of velocity with respect to distance. The fluid at the center of the pipe moves faster than the fluid adjacent to it, and so on to the surface, where the fluid velocity is zero. Thus the velocity gradient is dV/dr, where V is the x component of the velocity (Figure 3.6). This means that the fluid at the center has greater momentum (mV) than that close to the wall. In fluid mechanics it is convenient to talk of momentum per unit mass. For this mean flow that is all in the x direction, we say the x momentum varies as a function of r, or $V = V(r)$.

Now, an essential characteristic of fluid mechanics is that there is an interaction between the various parts of the fluid. What occurs in one region is felt in neighboring parts. If there is a velocity variation, it is passed on, usually modified in some way, to its neighbors. A convenient analogy is a large crowd of people moving toward, say, a sports ground. If some people speed up, those nearby must

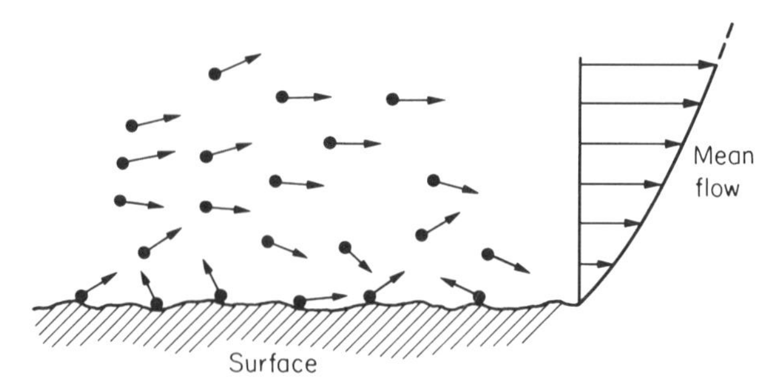

Figure 3.5 A simplified picture of the flow near a solid surface. Although far from the surface the molecules have a mean velocity (as well as a random component), very close to the surface this is reduced to zero because of the interaction of the molecules with the surface. This is a nominally smooth surface, but it must, because of the discrete nature of matter, have roughness at the smallest scales.

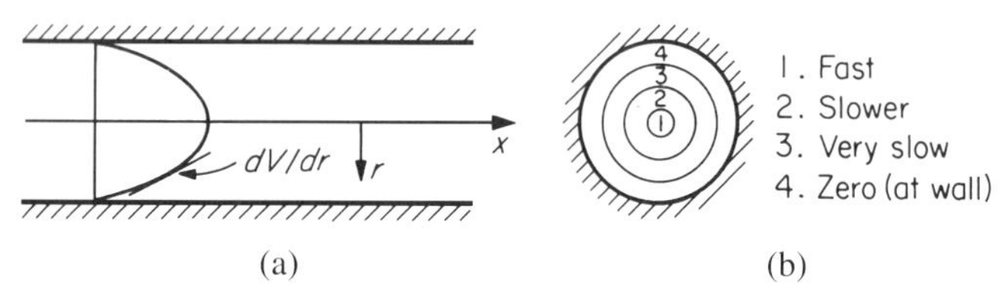

Figure 3.6 A closer look at laminar flow in a pipe. (a) Side view and (b) end view showing the annulae of fluid moving at different velocities. There is in reality a continuous variation of speed from a maximum in the middle to zero at the walls.

adjust to the change, and this propagates throughout the crowd. If the crowd is really large and we look at it from far enough away, it really does look like a fluid.

With this analogy in mind, let us return to our pipe flow (Figure 3.6). The fluid at the center of the pipe rubs past the slower fluid adjacent to it. The rubbing, or internal fluid friction, is achieved by the random motion of the molecules, which are, on the average, slightly faster for the fluid at the center of the pipe. By this means the fluid at the center of the pipe passes on some of its momentum to the fluid nearby, and that fluid in turn passes it on to the slower fluid next to it, and so on to the wall. Thus the x momentum is transferred from the center of the pipe toward the wall, from the region of high momentum to that of no momentum. We say the momentum is diffused down the gradient. We all are familiar with the diffusion of heat or matter down a gradient. For example, a patch of cream on the top of a glass of coffee diffuses down into the coffee. Here too the diffusion is caused by molecular interactions. The essential point is that cream diffuses from regions of high concentration to regions of low concentration. In the pipe, $V(x)$ momentum diffuses from regions of high $V(x)$ momentum to the region of low $V(x)$ momentum. The fact that momentum is a vector somewhat complicates the issue (x momentum is being transferred in the r direction), but the principle is similar to that of cream in coffee. Note that internal friction and diffusion of momentum are both due to the viscosity of the fluid. Both occur because of relative motion between various parts of the fluid. Both are due to the atomic nature of matter and both processes are irreversible. They are two sides of the same coin.

Just as for cream in coffee the rate of transfer of cream will be largest when there are large differences in concentration, so too for momentum the rate of momentum transfer will be greatest when the momentum gradient is the greatest. Because the momentum gradient is, if we consider the momentum per unit mass, the same as the velocity gradient, we can write

$$\left.\begin{array}{c}\text{The rate of transfer of}\\ V \quad \text{momentum}\end{array}\right\} \propto \frac{dV}{dr}. \tag{3.2}$$

From Newton's second law, a force must result from the momentum transfer, and thus Equation (3.2) can be written as

$$\text{Force} \propto \frac{dV}{dr}. \tag{3.3}$$

What is this force? The momentum of the fluid is transferred from the center of the pipe to the wall. Thus there must be a force acting on the wall itself. Now this force is along the pipe wall because it is caused by the fluid flowing in the x direction (Figure 3.6(a)). We call this type of force a shear force to distinguish from forces at right angles to surfaces, which are called normal forces. This shear force must be proportional to the area of the wall, and so it is convenient to talk about the shear force per unit area. This is called the **shear stress**. Its units are N/m^2. The velocity gradient, dV/dr, is often called the **shear**. Its units are s^{-1}. Thus,

$$\tau_{sh} \propto \frac{dV}{dr}, \tag{3.4}$$

where τ_{sh} is the shear stress along the wall caused by the transfer of momentum from the center of the fluid toward the wall. Equation (3.4) suggests that if a fluid is flowing in a pipe, the pipe will move unless it is restrained, because there is a force along the wall. Alternatively, if the pipe is restrained, then in order to move the fluid we must do work. We must exert a high pressure at one end of the pipe to overcome the stress at the wall that is caused by the transfer of momentum from the center of the pipe. We will show soon that this is the sole reason for the massive pumps we need to use in gas and water pipelines.

What is the constant of proportionality in Equation (3.4)? For exactly the same-sized pipe and exactly the same velocity at the center, will the stress at the wall be the same for water as for honey? In other words, will we have to push harder to move the honey? The answer, of course, is yes. It is the viscosity of the fluid that is the agent of momentum transfer, and Equation (3.4) becomes

$$\tau_{sh} = -\mu \frac{dV}{dr}. \tag{3.5}$$

I have included the minus sign in order to produce a positive stress, because in the pipe flow coordinate system V decreases as r increases (Figure 3.6(a)). In general, however,

$$\tau_{sh} = \mu \frac{dV}{dy}. \tag{3.6}$$

Here, y would, for example, be the distance from a solid surface (e.g., Figure 3.4(b) or (c)). Equation (3.5) (or (3.6)) states that the higher the viscosity, the higher the stress, and/or the higher the velocity gradient, the higher the stress. It makes intuitive sense: We have to work harder to push a fluid faster. We also have to work harder to push a highly viscous fluid in a pipe (or over any surface) than we do to push a fluid of lower viscosity at the same speed. Notice that Equations (3.5) and (3.6) imply that there are stresses anywhere in the fluid where there are velocity gradients; not just at the wall. Thus, apart from the pipe center line (where $dV/dr = 0$) each annulus of fluid exerts a stress on its neighbor (Figure 3.6). This stress (both within the fluid and at the wall), which is due to the relative motion of the adjacent layers (or laminae) of fluid, is in some ways analogous to the stress that occurs when you move your hand across a table surface. Here too there is relative motion, and the friction causes heating of both the hand and the table. For a fluid flow with velocity gradients there is also a temperature rise due to the internal friction, although it is often too small to notice. What is fundamentally different between motion in solids and fluids is that fluids cannot resist a shear stress, whereas solids can. This is often used as the fundamental distinction between solid and fluid mechanics.

The relation (3.5) (or (3.6)) provides some hint why turbulence occurs. As the flow in a pipe becomes faster and faster, the velocity gradient, and hence the stresses, get larger and larger because of the no-slip condition. At some point the laminar motion is unable to adequately transfer the momentum from the fast-flowing central region to the wall. In order to cope with the large momentum gradient, a breakdown occurs and the flow loses its order and becomes crinkled or turbulent. We will show later that the resultant turbulent eddies are well suited to transferring momentum from the center of the pipe to the walls. They act like large stirrers, providing rapid transfer of fluid from the central to the wall region. On the other hand, laminar flow is relatively inefficient in transporting momentum. Here the transporting agent is the molecular motion itself.

Equation (3.6) is a very basic relation. There are, however, two caveats. First, just as for the no-slip condition, peculiar fluids, such as paint, have a complex relation between stress and velocity gradient, and Equation (3.6) must be modified. We will not worry about these fluids here; after all, because all gases at moderate pressures (such as the atmosphere) as well as water, oil, and gasoline obey Equation (3.6), it is quite general enough for an introductory course! Second, when a fluid becomes turbulent, then although the no-slip condition at the wall still holds and thus there is a mean velocity gradient, there are, as we have observed, fluctuations in velocity, and therefore there must be local variations in the velocity gradient also. We will now turn to the modification of Equation (3.6) for turbulent flow.

3.2.2 Turbulent Flows

Look again at a smokestack (Figure 3.7), or at a cup of water being rapidly stirred (with a little dye for help in visualizing the flow), or at water rushing in a stream. The essential characteristic is its irregularity; nothing ever quite repeats itself, although on the average the general pattern is approximately the same. The turbulence appears to consist of many sizes of randomly rotating contorted eddy motions, the largest eddies being the size of the flow itself. The turbulent eddies are caused by the small relative differences in pressure and velocity that change rapidly throughout the fluid. Now, if there are velocity differences, there must be new stresses, and because the velocity fluctuations are irregular, these stresses must vary in direction and magnitude. Because the fluid is so contorted, there must be many of these new stresses. It is as if the internal structure of the fluid has become crinkled like an old person's skin, compared to a baby's, skin

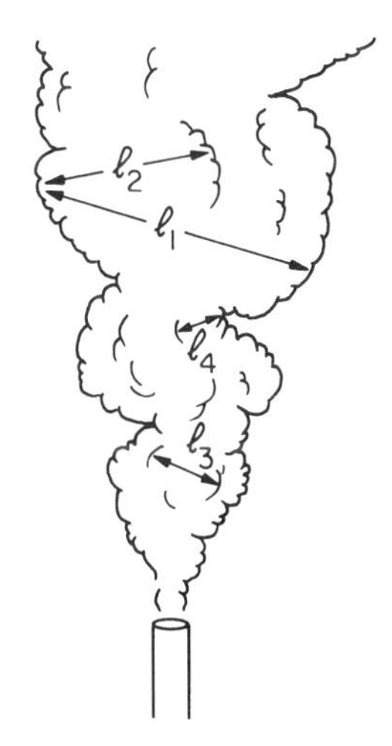

Figure 3.7 A smoke plume. Turbulence consists of eddies of various sizes. Because of this there are random velocity differences and hence random velocity gradients.

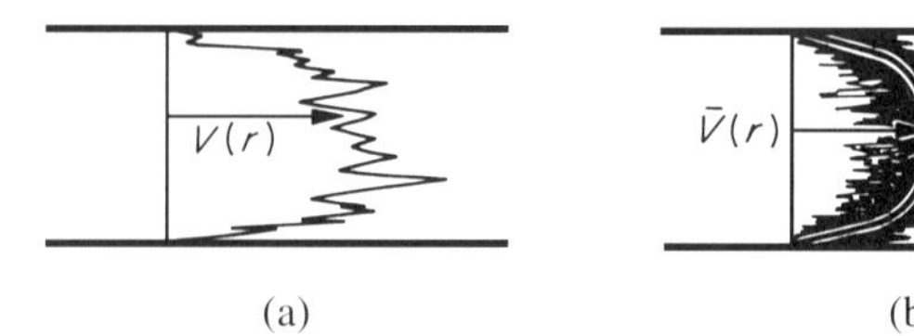

Figure 3.8 (a) The instantaneous velocity profile, $V(r)$, for turbulent flow in a pipe. (b) The average of many profiles shown in (a). This is called the mean velocity profile, $\overline{V}(r)$.

which is closer to the laminar situation. Crinkled skin has a far greater surface area than smooth skin, and so in turbulent motion the total area occupied by the shear stresses is far greater than for laminar flow. If the new random stresses in the turbulence canceled each other on the average, the dynamics of turbulence would not be much different from that of laminar flow. But this is not generally the case. If we look at turbulent flow in a pipe and measure the velocity across the pipe at one instant of time (by means of some fancy laser technique that can scan the whole flow very quickly), the profile will look something like that shown in Figure 3.8(a). If we superimpose hundreds, or thousands, of these profiles, a picture like Figure 3.8(b) will emerge. On the average there is a mean gradient just as for the laminar case, excepting here, instantaneously, there are departures from it. In both cases the no-slip condition must hold, so the velocity must be zero at the wall. (In fact, even for the turbulent case, very close to the wall there is a thin region (typically 1 mm or less) in which the flow is laminar. Here viscous forces dominate because of the very slow velocity. It is called the viscous sublayer.) Notice that the shape of the mean velocity profile for turbulent flow is different from the laminar profile (Figure 3.6). I will return to this in a moment.

Because there is, on the average, a greater velocity at the center of the pipe than near the wall, a turbulent **eddy**, characterized by the difference in velocity of its neighbors, will transport the fast moving fluid from the center of the pipe towards the walls and/or bring slower-moving fluid from the walls toward the center. It is as if the eddies are stirrers, transporting from regions of high average momentum (near the center of the pipe) to low momentum, at the wall. Although it is easy enough to see the eddies, for instance in a fast-flowing stream, mathematically characterizing them is difficult. We will say more about this in Sections 3.6 and 3.7. The combination of large-scale stirring (by the turbulence), coupled with the molecular transport down the instantaneous velocity gradients, makes turbulent flow very effective in transporting momentum from the center of the flow to the wall. In fact it is thousands and sometimes millions of times more effective than laminar flow. Greater transport of momentum means higher stresses at the wall, and this in turn means greater resistance to the flow. Great pipelines need pumps with power in the megawatt range because of the turbulence, which produces large stresses at the wall of the pipe.

It appears that Equations (3.5) and (3.6) must be modified for turbulent flows. But how? Because the stresses at the wall are higher, either $d\overline{V}/dr$ must be greater (where $\overline{V}$ is now the velocity averaged over a period of time (Figure 3.8)), or μ must change. Or perhaps the equation should have an extra term? In fact μ cannot change, because it is a property of the particular fluid: μ is the same for water whether it is turbulent or laminar, just as the density, ρ, is the same. What about $d\overline{V}/dr$? Notice that in Figure 3.8 I have drawn the velocity profile with a very sharp gradient indeed near the wall, implying a large stress at the wall. This is reasonable: Because turbulence transports momentum much more effectively than laminar flow does, it will even out the momentum as well as it can. Thus for most of the pipe, the velocity profile is quite flat compared to the (parabolic) laminar profile in Figure 3.6. But at the wall the no slip must still hold and so there is a dramatic gradient in velocity. Does this imply that away from the wall the transport of momentum is less efficient for turbulent flow because $d\overline{V}/dr$ is weaker? The answer is no because the eddies very effectively transport momentum throughout the flow. This suggests that Equation (3.6) needs an extra term. Because turbulence is random, this term must be of a statistical nature. Researchers still are unable to properly formulate it! There are very deep problems in understanding how the eddies interact with each other. After all, a careful examination of turbulence shows that there are various scales of motion. There are large, intermediate, and small eddies. These are all in random motion relative to each other. However, we do have some understanding of the way they interact, and this enables us to approximately determine the stresses at the wall of the pipe.

What engineers have learned to do is write empirical expressions for the stress at the wall of a pipe. Instead of attempting to write theoretical expressions, they do many experiments, varying the Reynolds number and tabulating the pressure drop as a function of Re. Formulas are derived from these experiments, but I will not bother you with them now. There is nothing very interesting to say about them. When needed I will supply them for solving problems. The essential point is that when the flow is laminar, we know a lot about it and we can determine the stress at the wall precisely. I will do this below in a problem. When the flow is turbulent, we must resort to approximate and empirical methods.

It is clear that turbulent flow is extremely important in engineering because of the large shear stresses it produces. This is, in general, a problem because it means more drag on cars and airplanes and more resistance in pipes. But just *because* turbulence causes greater stresses it has the ability to mix things much more rapidly than laminar flow. After all we have seen that wall stresses are the result of momentum transport. Turbulence not only transports momentum much more efficiently than does laminar flow, it can also transport heat, matter, chemicals, reactants, etc., much more efficiently, for exactly the same reason. Without turbulence, combustion in car and jet engines would occur too slowly to be worthwhile and smoke would hang around smokestacks and car exhausts for years! Life could not exist because we would be covered in the pollution from previous days, weeks, and years. Gently place some cream in your coffee. Watch how slowly it takes to mix. Now with one flick of your spoon, which makes the coffee turbulent, mixing is achieved in a few seconds.

Let us summarize our examination of the nature of laminar and turbulent flow.

All fluids are viscous, even air. Nearly all fluids (and certainly the ones most relevant to engineering, such as air, water, oil, and natural gas) exhibit the no-slip condition: their velocity at a solid surface, be it in a pipe, over an airplane wing, or over a blade of grass, is zero. We are referring here to the velocity parallel to the surface (Figure 3.4). (Of course the velocity normal to the surface must be zero unless there are holes it it!) As a result of the viscosity and the no-slip condition, there are stresses in fluids, causing resistance to motion, and therefore drag. When turbulence occurs at high Reynolds numbers, extra stresses occur because of the relative motion of the eddies. This greatly increases the drag or flow resistance. Although we have neat formulas for laminar flow (Equations (3.5) and (3.6)), we do not have simple, general relations for turbulent flow. This is somewhat frustrating yet at the same time provides a wonderful challenge.

3.2.3 The Calculation of Pressure Drop in a Pipe

We will end this section with an example that shows how to determine pressure drops in pipe flow. Once we have determined the pressure drop, it is a relatively straightforward procedure to determine the size of the pumps needed to overcome it.

Consider the flow of water in a pipe of length, $L = 100\,\text{m}$, and diameter, $d = 2.5\,\text{cm}$. The viscosity ν is $10^{-6}\ \text{m}^2/\text{s}$ and $\rho = 1{,}000\,\text{kg/m}^3$. Let us assume the Reynolds number is 2,000 and consider two cases: (a) the flow is still laminar and (b) the flow has become turbulent. (Remember, the transition from laminar to turbulent flow occurs at a Reynolds number of approximately 2,000. Therefore at this *Re* the flow could be either laminar or turbulent.) What is the total pressure drop in the pipe for each case? How much power is required to keep the water moving?

(a) Laminar Flow. Figure 3.9 shows a sketch of a pipe. We have isolated a small imaginary cylinder at the center of the pipe. There is a pressure drop dp,

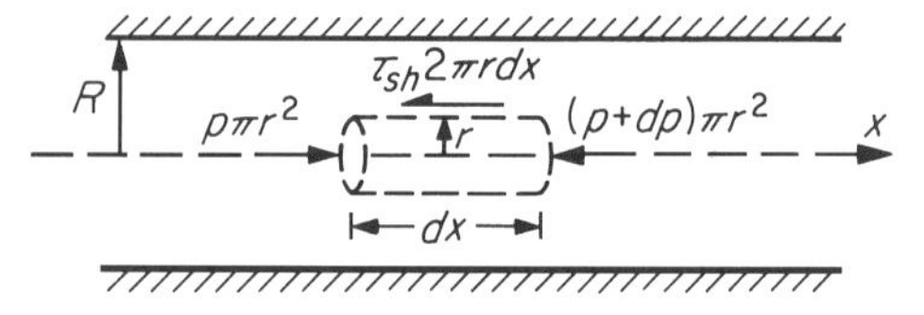

Figure 3.9 The forces on an imaginary cylinder in a moving fluid in a pipe. The flow may be laminar or turbulent, and is from left to right.

over the distance dx, and there is a stress all along the surface of the imaginary cylinder. Assuming the flow is in the positive x direction and the pressure increases with x (the signs will take care of themselves), a force balance shows

$$p\pi r^2 - (p + dp)\pi r^2 - 2\pi r dx \tau_{sf} = 0. \tag{3.7}$$

Here the first and the second terms are the normal forces on the left- and right-hand sides of the imaginary cylinder respectively, and the third term is the shear force on the surface of the cylinder. (There is no net normal force on the cylinder sides (by symmetry), and there are no shear forces on the ends because there is no radial velocity component.) Note that τ_{sf} is a force per unit area, so I have had to multiply it by the area of the cylinder (length dx and circumference $2\pi r$) to derive the force. Simplifying Equation (3.7), we find

$$-\frac{dp}{dx} = \frac{2}{r}\tau_{sf}. \tag{3.8}$$

Equation (3.8) shows that the pressure gradient along the pipe, $-dp/dx$, is determined by the stress; the larger the stress, the larger the pressure drop with pipe length, and thus the need for larger pumps.

Substituting Equation (3.5) into Equation (3.8), there results

$$\frac{dp}{dx} = \frac{2\mu}{r}\frac{dV}{dr}. \tag{3.9}$$

If the flow is uniform for the full length of the pipe, we can assume that $dp/dx =$ constant, i.e., over 1 meter the pressure drop will be twice what it was over 0.5 meter and so on. This allows us to solve the (differential) Equation (3.9) for the mean velocity profile. I will not do this, but will provide the answer:

$$V(r) = -\frac{R^2}{4\mu}\left(\frac{dp}{dx}\right)\left[1 - \frac{r^2}{R^2}\right], \tag{3.10}$$

where R is the pipe radius (Figure 3.9). If you differentiate (3.10) with respect to r and substitute it into Equation (3.9) you will see that it is indeed the solution. Notice the velocity profile, Equation (3.10), is a parabola with the maximum velocity at the center and zero velocity at the wall (Figure 3.10(a)).

Along the pipe center-line ($r = 0$), Equation (3.10) becomes

$$V_{\max} = -\frac{R^2}{4\mu}\left(\frac{dp}{dx}\right). \tag{3.11}$$

Now since we are assuming that the flow is uniform along the pipe, and thus that dp/dx is constant, Equation (3.11) implies that $V_{\max}$ is constant, i.e.,

$$-\frac{dp}{dx} = 4\mu V_{\max}/R^2 = \text{constant}. \tag{3.12}$$

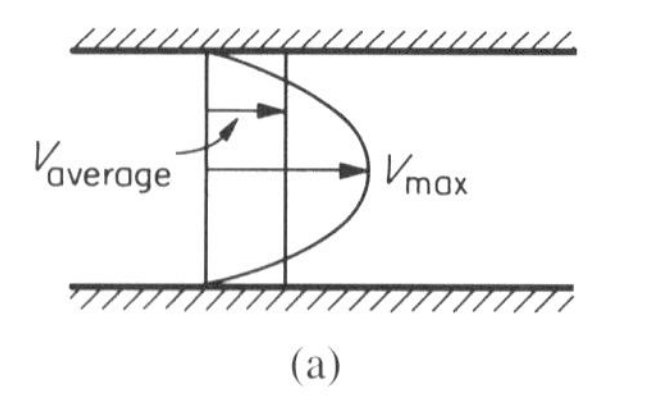

(a)

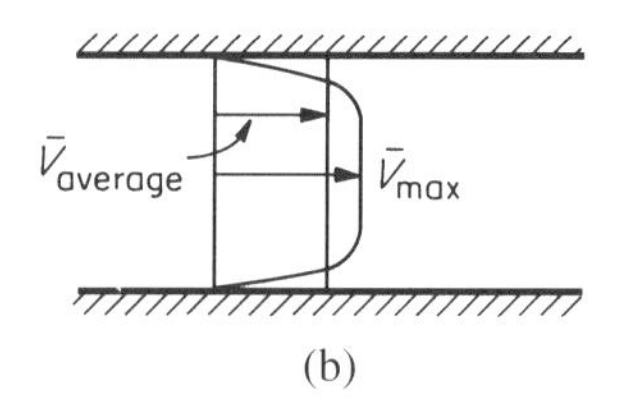

(b)

Figure 3.10 The velocity profiles for laminar (a) and turbulent (b) flows in a pipe. The average velocity is half that of the maximum for laminar flow. For turbulent flow it is around 0.8 times the maximum velocity.

Equation (3.12) may be integrated to yield

$$-(p_2 - p_1) = \Delta p = (4\mu V_{max}/R^2)(x_2 - x_1)$$
$$= 4\mu L V_{max}/R^2, \qquad (3.13)$$

where $x_2 - x_1 = L$ is the length of pipe for which the pressure drop, Δp, is being determined. Notice that p_2 is less than p_1, so $p_2 - p_1$ (and dp/dx) will be negative. Equation (3.13) shows that the pressure drop increases with increasing pipe length and this is consistent with our intuition. It also shows that decreasing the pipe diameter (but keeping μ, V_{max}, and L the same) will produce a greater pressure drop. Again this is reasonable: the velocity gradients (in the r direction) will be greater in the smaller pipe and the stresses will therefore be greater. This will give rise to greater pressure drop in the x direction.

Sometimes Equation (3.13) is written in terms of the *average* flow velocity, rather than in terms of its maximum value (at the center of the pipe). In problem 3.5 I ask you to show that the average flow velocity, V_{ave} (Figure 3.10) is $V_{max}/2$ for the laminar flow parabolic profile. So in terms of the average velocity, Equation (3.13) becomes

$$\Delta p = 8\mu L V_{ave}/R^2. \qquad (3.14)$$

This is very important design equation for laminar pipe flow since it provides a determination of the pressure drop for a given average flow speed, and fluid and pipe specification.

For the flow we are considering, the Reynolds number (based on the average flow velocity) is 2,000. Thus $V_{ave} = 2{,}000\ \nu/d = 2{,}000 \times 10^{-6}/(2.5 \times 10^{-2}) =$ 8 cm/s. From equation (3.14) we find $\Delta p = 8 \times 10^{-6} \times 1{,}000 \times 100 \times 8 \times 10^{-2}/(1.25 \times 10^{-2})^2 = 409.6$ Pa (here I have used the relation, $\mu = \rho\nu$). The power required to push this fluid is the normal force times the velocity of the fluid (think of a piston pushing the fluid). This is $\Delta p \pi R^2 V_{ave}$. For our example it is 1.6×10^{-2} W.

(b) Turbulent Flow. Here the force balance in Figure 3.9 still holds but we cannot use Equation (3.5) since the stresses are not related to the velocity gradient in such a simple way. As I have already mentioned, we do not have a general expression that relates the pressure drop to the shear for turbulent flows. However, engineers have developed empirical relations from experiments. They write their relations in terms of Reynolds number. One such relation is

$$\frac{dp}{dx} = -0.16(\rho V^2/d)Re^{-0.25}. \tag{3.15}$$

This is not exact, but is accurate to within a few percent for a smooth pipe. Since for a particular pipe flow all the terms on the right-hand side of Equation (3.15) are constant Equation (3.15) can be integrated in the same way we integrated Equation (3.12) to find $\Delta p = -0.16(\rho V^2/d)Re^{-0.25}L$, where L is the pipe length over which the pressure drop, Δp, is to be determined. For our example we find that for a Re of 2,000, $-dp/dx = 6.13$ Pa/m, or 613 Pa in 100 m. This is approximately 50 percent more than for the laminar flow at the same Reynolds number.

If we look at an example of say flow at 10 times the speed (0.8 m/s) in the same pipe, we find that for turbulent flow at this Re (20,000), the pressure drop (from Equation (3.15)) is 344 Pa/m or 3.44×10^4 Pa in 100 meters. This would require a power of 13.5 W to keep the flow moving. If the flow could, by some imaginary means, have been kept laminar up to this Re, then the pressure drop, from Equation (3.14), would have been 4.10×10^3 Pa in 100 m. This is nearly ten times less than for the turbulent flow. As the Reynolds number increases, the pressure drop due to turbulence increases dramatically when compared to what the pressure drop would have been if the flow could have been kept laminar. For this reason engineers spend much time trying to counter the effects of turbulence: such research is called drag reduction. Notice that the pressure drop in our relatively short and low volume-flow-rate pipe has been small. Problem 3.6 shows that in large pipes, the pressure drop is significant, and so too are the power requirements for pumps.

3.3 The Bernoulli Equation

We have shown that fluid flowing in a pipe, be it air or water or any other fluid, will experience a pressure drop due to friction. However, the pressure may also change because of elevation changes and changes in the pipe area. In this section we will develop an equation that takes these aspects into account. In order to simplify matters, we will assume that friction losses are negligible. As we have already shown, this can never be the case; the viscous stresses always lead to pressure losses. Nevertheless, for some problems these losses may be small compared to pressure changes that occur due to changes in velocity or elevation of the fluid. When friction can be neglected, we call the fluid motion **ideal** or **inviscid** flow. We will study this type of motion here, and then return, in the next section, to the cases of fluid flow for which friction is important.

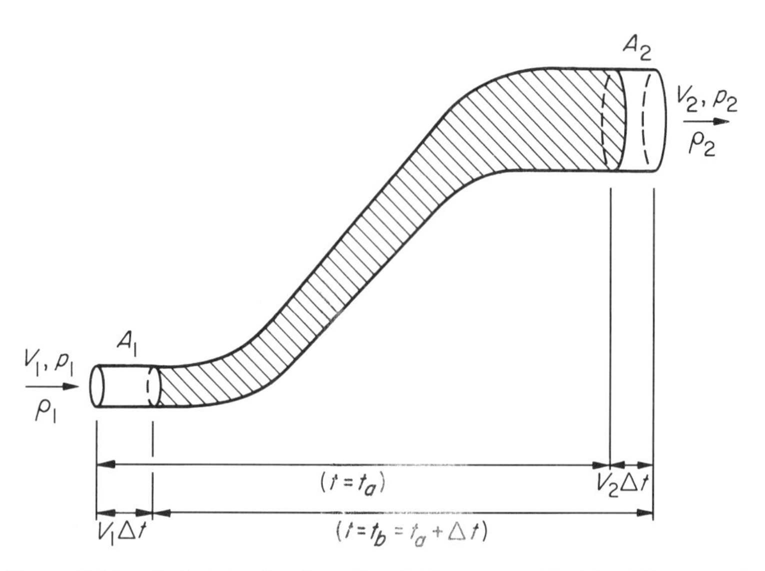

Figure 3.11 A sketch of a pipe of variable area and height. The control volume at time $t = t_a$ is the shaded region plus the small region to its left. At $t = t_b$ the control volume is the shaded region plus the small region to its right. Note that $V_1 \Delta t \neq V_2 \Delta t$ since the pipe area changes.

Instead of the simple constant-area horizontal pipe flow of Figures 3.6 or 3.8, let us consider the more general case of a pipe that varies in area and goes up an incline, as shown in Figure 3.11. A section of the pipe is isolated at time $t = t_a$, and a new section is isolated a moment Δt later at time $t_b = t_a + \Delta t$. The area of the pipe is A_1 where the fluid is entering the region of interest and A_2 where it is leaving. Don't worry that these two areas may change a little from times t_a to t_b; this change can be made vanishingly small as $\Delta t \to 0$. Let the fluid velocity and density respectively be V_1 and ρ_1 (entering) and V_2 and ρ_2 (leaving). We know from Section 3.2 that the fluid velocity varies over the area of the pipe; here V_1 and V_2 are the average velocities (Figure 3.10). Notice that after the short time interval the fluid has moved $V_1 \Delta t$ at section 1 and $V_2 \Delta t$ at Section 2.

Now, if our pipe is not leaking, the mass going in during time Δt must equal the mass leaving. The mass of fluid entering section 1 during time Δt is the differential volume of the pipe multiplied by the fluid density, that is, $A_1 V_1 \Delta t \rho_1$, and so the conservation of mass states

$$A_1 V_1 \rho_1 \Delta t = A_2 V_2 \rho_2 \Delta t \qquad \text{or} \qquad A_1 \rho_1 V_1 = A_2 \rho_2 V_2. \tag{3.16}$$

This expression must hold everywhere in the pipe, not only at the ends. Sometimes it is written in the form

$$A\rho V = \text{constant}. \tag{3.17}$$

The product $A\rho V$ is a mass per unit time, or **mass flow rate**, dm/dt. We will often use the shorthand notation $dm/dt \equiv \dot{m}$. Equation (3.17) states that the mass flow rate is constant. Notice that if ρ does not vary then as the area increases the average speed decreases. This is intuitive. However, if the density decreases (say, for example, if the fluid is a gas that is being heated in the pipe), Equations (3.16) or (3.17) show that if A increases, V may in fact increase also, depending on the actual values. If ρ = constant the flow is called **incompressible**. Most liquids such as oil or water can be considered incompressible, particularly for pipe flow problems. Equations (3.16) or (3.17) do not hold if the flow is **unsteady** – if, for instance, the flow into the pipe is suddenly increased or decreased. In this chapter we are going to consider only **steady flow**.

We are interested in the pressure change in the pipe. To determine it we will estimate the work done in moving the fluid during the time interval Δt. The work done on the fluid by the fluid entering at section 1 is $\mathbf{F} \cdot d\mathbf{x} = p_1 A_1 V_1 \Delta t$. This is the work required to push against the fluid already in the control volume. Similarly, the work done by the exiting fluid on the fluid downstream from it is $p_2 A_2 V_2 \Delta t$. Thus the net work done on the fluid between sections 1 and 2 is

$$p_1 A_1 V_1 \Delta t - p_2 A_2 V_2 \Delta t.$$

Notice the minus sign because at section 2 work is being done on the surroundings (the downstream fluid). We are back to the language of thermodynamics, with the system now the length of pipe. Note that the fluid in the system remains the same apart from the small amount entering and leaving ($A_1 V_1 \rho_1 \Delta t$ and $A_2 V_2 \rho_2 \Delta t$ respectively), and this corresponds to the shape of the system changing from that shown by the arrow, $t = t_a$, to that shown by the arrow, $t = t_b$.

From the first law, the work done on the system must equal the change in energy of the system (we assume for the moment there is no heat transfer), and thus

$$p_1 A_1 V_1 \Delta t - p_2 A_2 V_2 \Delta t = \Delta E, \tag{3.18}$$

where ΔE is the total change of energy of the system. Now, there is no change of energy for the fluid in the shaded region because after time Δt the potential energy, kinetic energy, and internal energy are exactly the same as they were originally. The fluid at any point has been replaced with fluid having exactly the same energy (remember that the flow is steady). On the other hand, the energy of the fluid entering will be in general different from the energy of the fluid leaving the system. Thus the change in energy of the system, ΔE, will be

$$m_2\left(\frac{1}{2}V_2^2 + gz_2 + u_2\right) - m_1\left(\frac{1}{2}V_1^2 + gz_1 + u_1\right),$$

where $\frac{1}{2}V^2$, gz, and u are the kinetic, potential, and internal energy per unit mass respectively and the subscripts refer to the entering and leaving fluid. (We must have a reference for z. Any height will do (such as the height of the pipe at position 1) because we are interested in differences only.) Now, $m_1 = m_2 = m$ (the mass entering must equal the mass leaving), so Equation (3.18) becomes

$$p_1 A_1 V_1 \Delta t - p_2 A_2 V_2 \Delta t = m\left(\frac{1}{2}V_2^2 + gz_2 + u_2 - \frac{1}{2}V_1^2 - gz_1 - u_1\right). \tag{3.19}$$

If we divide Equation (3.19) by m and simplify by noting that $A_1 V_1 \Delta t/m = 1/\rho_1$ and $A_2 V_2 \Delta t/m = 1/\rho_2$ (Equation (3.16)), there results

$$\frac{p_1}{\rho_1} + \frac{1}{2}V_1^2 + gz_1 + u_1 = \frac{p_2}{\rho_2} + \frac{1}{2}V_2^2 + gz_2 + u_2. \tag{3.20}$$

This very important equation is known as the **Bernoulli equation**. Because positions 1 and 2 were chosen arbitrarily, we can write Equation (3.20) as

$$\frac{p}{\rho} + \frac{1}{2}V^2 + gz + u = \text{constant}. \tag{3.21}$$

This is the more usual way of writing the Bernoulli equation.

Because we have assumed that the fluid is frictionless, we know that the Bernoulli equation can never hold exactly. Yet it is a very powerful equation because it relates pressure changes to velocity changes. It is widely used in engineering analysis. Although we have developed it for flow in a pipe, its application is more general than that. For other kinds of fluid motion we can imagine there are pipes, or **streamtubes**, bounding a particular region of the flow. They are defined so that no fluid can cross their surface (Figure 3.12).

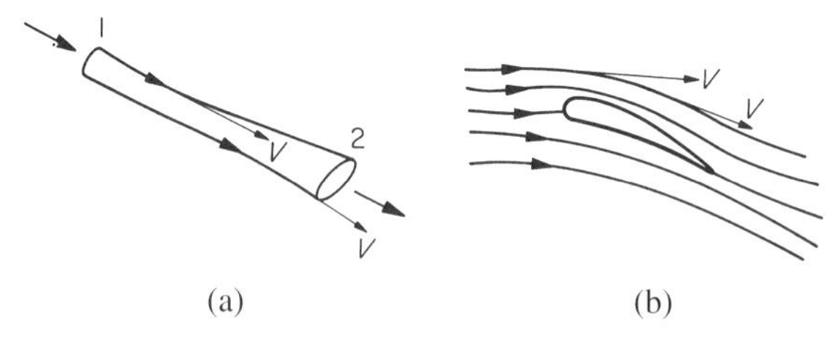

Figure 3.12 Streamlines and streamtubes. The flow velocity is always tangent to a streamline or streamtube, so no fluid crosses them. (a) A stream tube. (b) Streamlines around a wing.

The lines making up the streamtube are called **streamlines**. The velocity vector must always be tangent to a steamline because no fluid can cross the boundary of the streamtube. So if the velocity field is known, the streamtubes can be determined. Using this approach the pressure change over an airplane wing, or in the jet stream of the atmosphere, can be estimated by considering the flow along a streamline.

For incompressible flow ($\rho = \text{constant}$) the internal energy of the fluid cannot be changed because the density of the fluid cannot be affected by changes of pressure. (We recall that because by assumption the fluid is inviscid, the internal energy cannot be changed because of internal friction either, because there is none.) For this case the Bernoulli equation (3.21) reduces to

$$\frac{p}{\rho} + \frac{1}{2}V^2 + gz = \text{constant}. \tag{3.22}$$

Liquids (water, oil, etc.) are incompressible. But for slow flow in a horizontal pipe or duct, gases such as air can be considered to be incompressible too.

We will explore the consequences of Equation (3.22) in the next section. Here we note that for $V = 0$ (i.e., for a fluid at rest) it becomes $p/\rho + gz = \text{constant}$, or

$$p_1 - p_2 = \rho g(z_2 - z_1). \tag{3.23}$$

Thus for an incompressible fluid, such as the ocean, the pressure increases linearly with depth. Equation (3.23) is known as the hydrostatic condition for an incompressible fluid. We will derive it in a different way in Section 5.1.4 where we will extend it to a compressible fluid, such as the atmosphere.

3.3.1 Some Applications of the Bernoulli Equation

We will now apply the Bernoulli equation to some particular flows. In all cases we will assume the fluid density is constant, so we will use Equation (3.22).

In many engineering problems fluids discharge from vessels. This is shown in Figure 3.13(a) where some streamlines are drawn for the fluid flowing from the high-pressure, p_1, to a low-pressure region, p_2. We will assume at discharge that the pressure of the fluid is the same as that of its surroundings. In the vessel, far to the left of the orifice, the fluid speed will be negligible compared to that of the orifice itself, so $V_1 = 0$. (We are assuming the vessel is large, so the fluid is close to stagnant.) We can now write the Bernoulli equation for flow along the central streamline. Here there is no change in height, so Equation (3.22) becomes

$$\frac{p_2}{\rho} + \frac{V_2^2}{2} = \frac{p_1}{\rho} + 0.$$

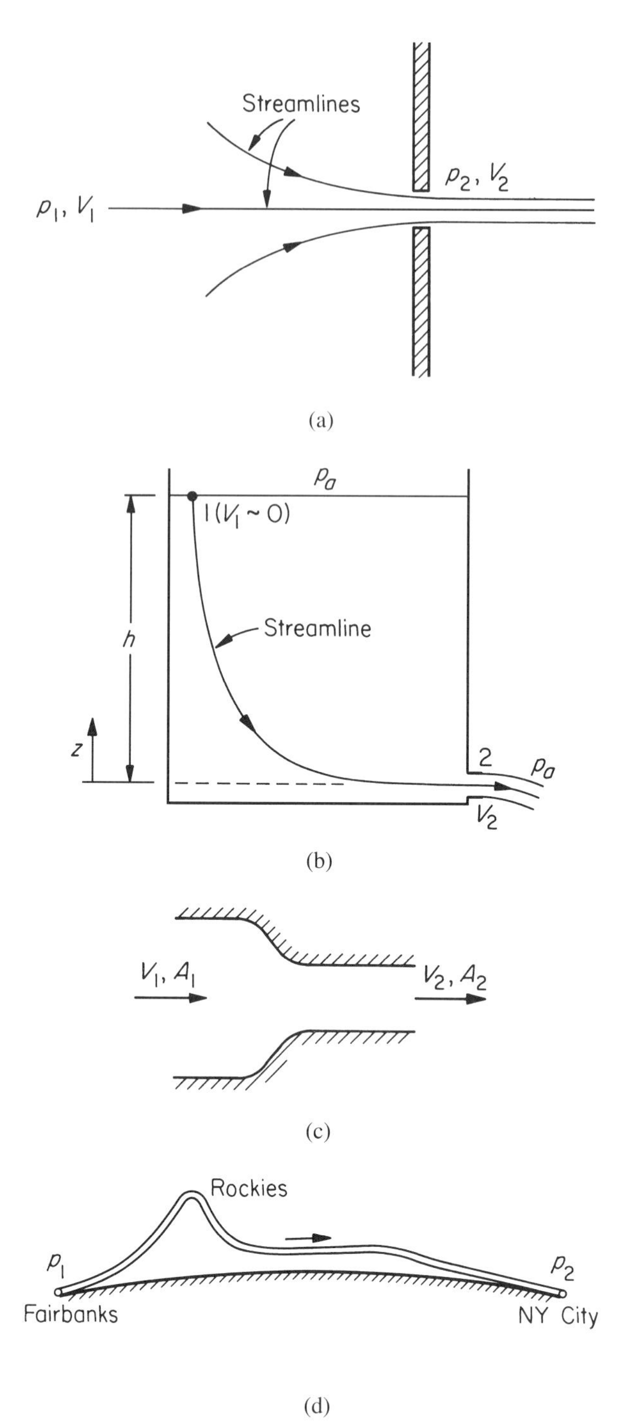

Figure 3.13 (a) Flow from a high-pressure vessel (p_1), through an orifice, to a region of low pressure (p_2). A few streamlines are shown. The fluid is essentially at rest ($V_1 \sim 0$) far to the left of the orifice. (b) Water discharging from a vessel. (c) Flow through a wind tunnel contraction. (d) A constant-area pipeline.

Thus we find,

$$V_2 = \sqrt{\frac{2(p_1 - p_2)}{\rho}}. \tag{3.24}$$

From this we can determine the mass flow rate, $\dot{m}(= \rho VA)$ of the fluid discharging from the vessel to be

$$\dot{m} = A\sqrt{2\rho(p_1 - p_2)}, \tag{3.25}$$

where A is the area of the orifice. As we would expect, the higher the fluid pressure, the faster the flow rate.

The high pressure in the vessel could be caused in many ways. But consider it to be just the result of the mass of fluid above it. For example, the fluid in the vessel could be water, discharging into air. Figure 3.13(b) shows the situation. We can apply the Bernoulli equation from the top of the vessel to the orifice. It becomes

$$\frac{p_a}{\rho} + 0 + gh = \frac{p_a}{\rho} + \frac{V_2^2}{2} + 0,$$

that is,

$$V_2 = \sqrt{2gh}. \tag{3.26}$$

So we find that the speed of the fluid at the orifice is the same as that of an object falling through the same distance, h. (In fact h will decrease slowly with time, making this into an unsteady problem. Here we are assuming a very large tank ($V_1 \sim 0$) and small times, so h is approximately constant.)

We now turn to wind tunnel design. Figure 3.13(c) shows a test section of a wind tunnel. Before the test section there is an area of large cross section called the settling, or plenum, chamber. Here the flow velocity is slow, so that any disturbances in the stream can settle down. After passing through the plenum, the fluid then is accelerated through the contraction, into the test section. Again we will assume that the flow is incompressible (a very good assumption for a water tunnel, and also good for air at speeds much less than the speed of sound (340 m/s), that is, for speeds less than about 50 m/s (approximately 100 mph)). Neglecting any height changes, the Bernoulli equation (3.22) and the conservation of mass (Equation (3.16)) become

$$\frac{p_1}{\rho} + \frac{1}{2}V_1^2 = \frac{p_2}{\rho} + \frac{1}{2}V_2^2 \tag{3.27}$$

and

$$V_1 A_1 = V_2 A_2. \tag{3.28}$$

Substituting Equation (3.27) into Equation (3.28) we find

$$p_2 - p_1 = \Delta p = \frac{1}{2}\rho V_2^2\left[\left(\frac{A_2}{A_1}\right)^2 - 1\right]. \tag{3.29}$$

Because $A_2 < A_1$ (Figure 3.13(c)), then Δp is negative. When the fluid speeds up, its pressure drops. Intuition may suggest the opposite. The reason for the pressure drop is that work must be done to accelerate the fluid through the contraction. Something must be forfeited, and it is the pressure within the fluid. Notice that Equation (3.29) may be used to determine the pressure change in a diffuser, where there is an area increase rather than decrease. (Reverse the direction of the flow in Figure 3.13(c).) Here Δp will be positive. As the fluid decelerates, it gains pressure. If we place a diffuser (of the same area ratio) after a contraction, the pressure will be recovered if there are no losses due to friction.

In the above examples we have assumed ideal or inviscid flow. As we have already discussed (Section 3.2), there is really no such thing as ideal flow, because all fluids have viscosity. In order to show that fluid friction is sometimes negligible, such as for the contraction, we will reconsider the laminar pipe flow problem (Section 3.2.3) where we showed the pressure drop was 409.6 Pa in 100 m or 4.096 Pa/m for an average flow speed of 8 cm/s in a 2.5-cm-diameter pipe. If this flow entered a contraction, reducing its diameter, say, from 2.5 to 0.5 cm, Equation (3.29) shows that $p_2 - p_1 = -2{,}000$ Pa. Here we have taken $\rho = 1{,}000\,\mathrm{kg/m^3}$ and used the conservation of mass (Equation (3.13)) to determine V_2 from V_1 (which was 8 cm/s). Thus the contraction has reduced the flow pressure by nearly 500 times more than the pressure drop due to friction for 1 m of pipe length. Clearly, for this example the pressure drop due to friction is unimportant compared to the pressure drop due to the area change.

Yet if we consider long pipes, friction *will* become important. We conclude this section with a return to constant-area horizontal pipe flow. If we assume the density is constant (e.g., the fluid is water or oil), then because A is constant the conservation of mass (Equation (3.16)) shows that V is constant. For this case the Bernoulli equation (3.22) shows that $p =$ constant. Assuming the fluid is ideal, then, leads to an absurdity for long lengths of pipe. For example, if the pipe started in Fairbanks, Alaska, and ended in New York City (Figure 3.13(d)), and if its area did not change, and if it began and ended at sea level, then the ideal fluid assumption indicates there will be no pressure drop and therefore no need for pumps. Once started, the fluid would flow on its own accord. In fact many megawatts of power are needed to pump fluids along large pipelines, as you will see from Problem 3.6. Clearly the inviscid fluid assumption cannot be used. A large pressure drop will be caused because of the fluid friction. This is why we spent so much time discussing fluid friction in Section 3.2. This example indicates that the Bernoulli equation must be used with caution.

3.4 A Further Look at the First Law of Thermodynamics: Open Systems

The Bernoulli equation (3.20 or 3.21) states that for an inviscid fluid, the sum of the kinetic energy, potential energy, and internal energy (all per unit mass), plus the pressure divided by the density, is constant from position to position along a pipe. All terms in the Bernoulli equation have the units of energy per unit mass, and thus the equation is an energy balance, or statement of energy conservation. In fact, the Bernoulli equation is a particular form of the first law of thermodynamics for an **open system**: one in which mass can cross the boundaries (as opposed to the closed system of Chapter 2 in which no mass crossed the boundaries). In order to make the first law for an open system completely general we must consider work and heat interactions, just as we did for closed systems in Chapter 2. We must also take friction losses into account. That is the purpose of this section.

Before we begin, however, it is important to distinguish between the form of the pressure term in the first law for a closed system (pdv work, Equation (2.23)) and the pressure term appearing in the Bernoulli equation (Equation (3.20)). The pressure term appears in the Bernoulli equation because work must be done on (or by) the fluid in the control volume by the fluid entering (or leaving it), in order to move the flow. Otherwise its motion could cease because of the decreases in kinetic, potential, or internal energy. This pressure term appears whether or not the density is constant. On the other hand, for the closed system, pdv work occurred only when the fluid was changing volume and thereby changing the density of the fluid in the system (because it is closed its mass remains constant). Of course the basic physical concept of pressure is the same in both open and closed systems: It is the force per unit area exerted normal to any plane within the fluid.

We will now extend the Bernoulli equation to include heat and work interactions and flow losses. In Figure 3.14, we have redrawn the pipe of Figure 3.11 but have allowed heat to cross the pipe walls. We have taken the heat direction as going into the fluid; for instance, the pipe may be passing through a hot boiler tank or it may be wrapped around with a heating coil, but the direction could be the other way, such as occurs with a hot radiator pipe in a room. If the direction

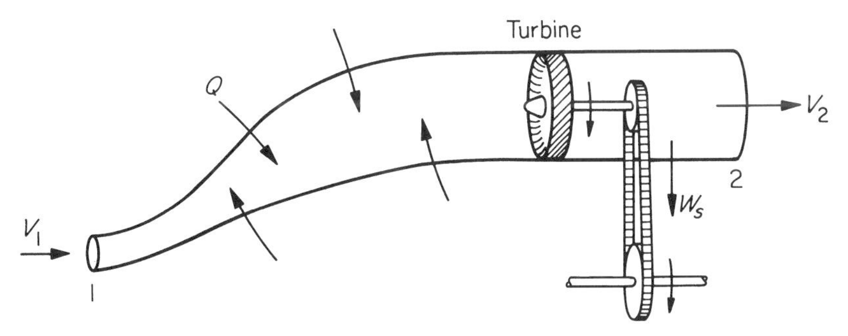

Figure 3.14 An open system with heat and shaft work interactions across its boundaries.

of the heat interaction is into the pipe, the fluid will gain energy; if it is out of the pipe, the fluid will lose energy.

We have also drawn a propeller in the pipe of Figure 3.14. This propeller is on a shaft that is coupled to the outside of the pipe by means of a pulley or mechanical coupling. If the propeller is driven from the outside by means of a motor, it will put energy into the system. **Shaft work** will be done on the system. This could keep the fluid flowing, and for this case the propeller would be called a *pump* or *compressor*. On the other hand, the fluid could drive the propeller. Here the fluid is doing work on the surroundings, and this could be used to drive a power generator. When a propeller is used in this way it is called a *turbine*. Although the actual shapes of the blades on pumps and turbines will be different, and they will also depend on the type of fluid, the principle remains the same. They are both devices that change the energy of the fluid by means of work interactions.

Because the heat interaction, the shaft work interaction, and the fluid friction all change the energy of the system, we must generalize the Bernoulli equation to include them. Equation (3.20) will now become

$$q + w_s + w_f = \left(\frac{p_2}{\rho_2} + \frac{1}{2}V_2^2 + gz_2 + u_2\right) - \left(\frac{p_1}{\rho_1} + \frac{1}{2}V_1^2 + gz_1 + u_1\right). \quad (3.30)$$

We use the subscript s to denote shaft work due to the pump or turbine. We have denoted the flow losses due to friction by the symbol w_f. The symbol for work is used because work has to be done by the fluid to overcome these losses. Unlike w_s, the sign of w_f can only be negative. We have shown how to determine its value for flow in a pipe in Section 3.2. Of course, when $q = w_s = w_f = 0$, Equation (3.30) reduces to Equation (3.20).

The group $p/\rho + u$ or $pv + u$ often occurs in thermodynamics and is used to define a new property called the **enthalpy** (see Problem 2.8):

$$h \equiv pv + u. \quad (3.31)$$

Because p, v, and u are measurable properties of a fluid, so is h. Just as we can talk of, say, a region of water having a particular temperature or pressure, so can we talk of its enthalpy, which has units of J/kg.

Using Equation (3.31), Equation (3.30) may now be written in the form

$$q + w_s + w_f = \left(h_2 + \frac{1}{2}V_2^2 + gz_2\right) - \left(h_1 + \frac{1}{2}V_1^2 + gz_1\right). \quad (3.32)$$

It is often useful, for open systems, to multiply Equation (3.30) or Equation

(3.32) by $\dot{m}$, the mass flow rate:

$$\dot{Q} + \dot{W}_s + \dot{W}_f = \left(\frac{p_2}{\rho_2} + \frac{1}{2}V_2^2 + gz_2 + u_2\right)\dot{m} - \left(\frac{p_1}{\rho_1} + \frac{1}{2}V_1^2 + gz_1 + u_1\right)\dot{m} = \left(h_2 + \frac{1}{2}V_2^2 + gz_2\right)\dot{m} - \left(h_1 + \frac{1}{2}V_1^2 + gz_1\right)\dot{m}. \quad (3.33)$$

Here $\dot{Q} = q\dot{m}$ and $\dot{W} = w\dot{m}$. The units of all the terms in Equation (3.33) are watts (W).

Equations (3.32) and (3.33) allow us to determine the heat and work transfer to and from open systems if we know the energy changes internal to the system. Or, given design requirements of shaft work or heat transfer, they allow us to determine the energy changes. We will now apply them to the analysis of gas turbines and jet engines.

3.5 Gas Turbines and Jet Engines

Engines and their relation to the environment provide the principal focus of this book. In Chapter 2 we analyzed IC piston-cylinder engines in terms of a closed thermodynamic system. Although nearly all cars are driven by IC engines, the power they produce is small compared to our industrial and domestic needs. For example the engine in a mid-sized car produces around 150 kW (200 horsepower), enough to move a few tons of car plus occupants but not enough to provide power to more than a handful of houses. Large piston engines have been used to propel ships and for domestic and industrial power generation, but as they are scaled up their efficiency decreases. So, in order to generate large amounts of power, a completely different approach is used: Instead of a closed system, (e.g., Figure 2.7), fluid is made to flow past a turbine, which produces work. Here the system is open, and its analysis must be in terms of the first law for an open system (Equation (3.32) or (3.33)).

As for IC engines, open power systems must undergo a cycle and energy must be supplied in the form of heat. Steam power plants using this principle have been in use for over 100 years. Here we will describe a power plant that uses air. Such a system is called a gas turbine power cycle. It is used in some cars but mainly in jet engines and more recently for power generation. The principle is simple: Air passes through a turbine, turning it so that it can produce work by means of some coupling to the surroundings (Figure 3.15). If, when the air passes through the turbine, the change in kinetic energy of the air stream is small (the duct area is approximately constant), and assuming the pipe is insulated (adiabatic flow)

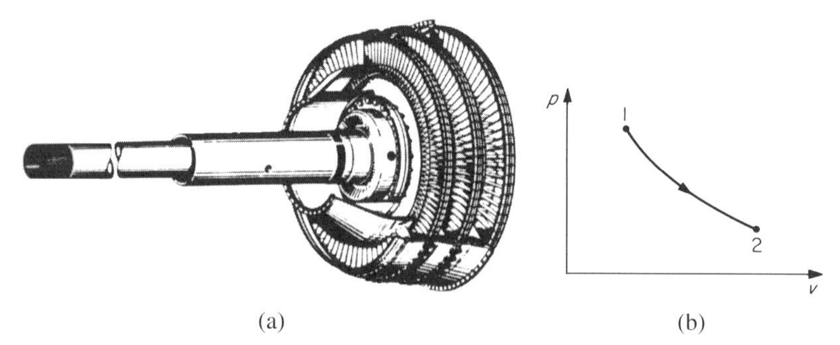

Figure 3.15 A gas turbine (a) and its p-v diagram (b).

and that friction losses are negligible, then Equation (3.33) shows

$$\begin{aligned}\dot{W}_s &= (p_2 v_2 - p_1 v_1 + u_2 - u_1)\dot{m} \\ &= (h_2 - h_1)\dot{m}.\end{aligned} \tag{3.34}$$

Note that $\dot{W}_s$ will be negative because the system is doing work on the surroundings. So, $h_2 < h_1$. We see that the shaft work results in a decrease in enthalpy. Now, if we assume that the air in the turbine can be described as a perfect gas (in fact there will be some departure due to the high pressures, but we will neglect this), then it follows that $h = pv + u = RT + u$. For a change in T and u we may write this as $\Delta h = R\Delta T + \Delta u = (R + c_v)\Delta T$. In Problem 2.8 you determined for a perfect gas that $c_p - c_v = R$. Thus we find that $\Delta h = h_2 - h_1 = c_p(T_2 - T_1)$. So for a perfect gas the enthalpy change is proportional to the temperature change. Hence it is evident that the temperature of the gas must decrease when passing though the turbine. From this it follows, that for a given outlet temperature, T_2, the greater the inlet gas temperature, the greater will be the shaft work produced by the turbine. If we had not neglected internal friction, then some of the enthalpy (or temperature) would have to be used in overcoming the losses, and thus for the same ΔT, less shaft work would be produced. In practice most of the losses are produced by the turbine blades. However, compared to the energy produced, the losses are small.

In Chapter 2 we found that for adiabatic processes (such as we are assuming for our turbine), the temperature, pressure, and specific volume obey the following relations (Equations (2.43) and (2.44)):

$$\left(\frac{T_1}{T_2}\right)^{c_v/R} = \frac{v_2}{v_1} \quad \text{and} \quad \frac{p_1}{p_2} = \left(\frac{v_2}{v_1}\right)^{(R/c_v+1)}.$$

Here state 1 is the entrance to the turbine and state 2 is the exit (Figure 3.15). Although these relations were developed for a closed system, they can also be applied to open systems because they only relate changes in properties undergoing a prescribed (adiabatic) path. On the other hand, the work done will be different because the first law is quite different for closed and open systems. Equations (2.43) and (2.44) can be combined to yield

$$\frac{T_1}{T_2} = \left(\frac{p_1}{p_2}\right)^{1/(1+c_v/R)} = \left(\frac{p_1}{p_2}\right)^{R/c_p}. \tag{3.35}$$

For air $T_1/T_2 \sim (p_1/p_2)^{0.286}$. Because the temperature decreases across the turbine, Equation (3.35) shows that so also must the pressure. We see then that the enthalpy drop across a turbine results in both a temperature and a pressure drop. The perfect gas law then indicates there must be an increase in v. The path is shown on the p-v diagram of Figure 3.15(b). The gas turbine should be contrasted with an (incompressible) water turbine ($\rho = \text{constant}$). Here, if there is no friction ($u_1 = u_2$), Equation (3.33) shows that the shaft work is due solely to the pressure drop, whereas in a gas turbine the enthalpy change results from the temperature change, and this will occur even if friction losses can be neglected. In real gas or water turbines there will be considerable friction losses, so a higher enthalpy (or pressure) drop will be needed to produce the required shaft work.

Clearly, in order to build a working gas turbine, we must produce a gas with high pressures and temperatures so it can be discharged through the turbine to produce the required shaft work. The high temperatures are produced in a combustion chamber, where heat is usually added to the flowing gas at a constant pressure. In order to produce the high pressure, the gas is first compressed. A sketch of the system is shown in Figure 3.16(a). In order to complete the cycle, we have let the air from the turbine pass through a **heat exchanger**. Here the air is cooled at a constant pressure until it reaches state 1, to begin its cycle again. Notice we have changed the numbering of the various positions in the cycle compared with Figure 3.15. Here the turbine is between 3 and 4. The

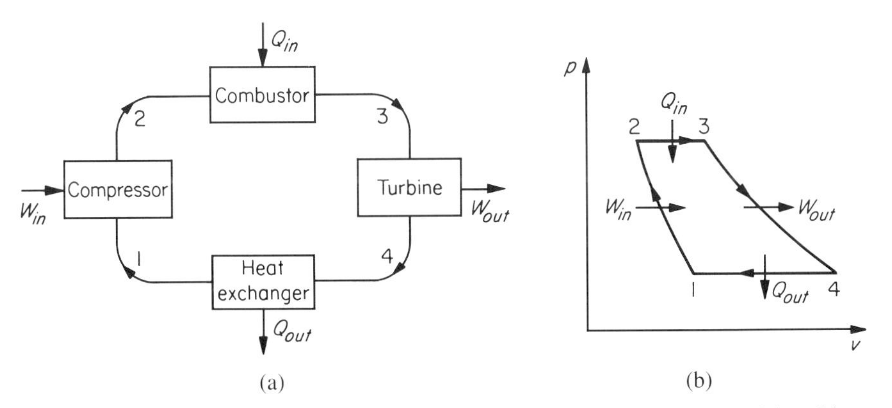

Figure 3.16 The components of a gas turbine power plant system (a) and its p-v diagram (b). This is known as the ideal Brayton cycle.

gas turbine system of Figure 3.16(a) is now closed: There are only heat and work interactions across its boundaries. Air first goes through the **compressor** where the pressure (and hence enthalpy, Equation (3.32)) is increased by means of shaft work done on the system. If this is done adiabatically, the path is of the same form as for the turbine but in the opposite direction (Figure 3.16(b)). After going through the combustor, where the fuel is burned, the hot high-pressure air discharges through the turbine to produce work. Here, as for any cycle, the efficiency is defined as in Equation (2.51):

$$\eta_{\text{th}} = \frac{|W_{\text{net}}|}{Q_{\text{in}}} = \frac{|W_{\text{turbine}}| - |W_{\text{pump}}|}{Q_{\text{in}}}. \tag{3.36}$$

Assuming negligible friction losses, kinetic energy changes and height changes, the first law (Equation 3.32) applied to each term of (3.36) then shows

$$\eta_{\text{th}} = \frac{-(h_4 - h_3) - (h_2 - h_1)}{h_3 - h_2}. \tag{3.37}$$

Notice the minus sign. This is included so that the turbine work is positive, in accordance with our definition of efficiency.

For a perfect gas $dh = c_p dT$. Thus, if c_p is constant, Equation (3.37) can be written as

$$\eta_{\text{th}} = \frac{-(T_4 - T_3) - (T_2 - T_1)}{T_3 - T_2}. \tag{3.38}$$

It is convenient to rearrange Equation (3.38) in the form

$$\eta_{\text{th}} = \frac{-T_3[T_4/T_3 - 1] + T_2[T_1/T_2 - 1]}{T_3 - T_2}. \tag{3.39}$$

Paths 1–2 and 3–4 are adiabatic, so we may use Equation (3.35) to convert temperature ratios into pressure ratios:

$$\eta_{\text{th}} = \frac{-T_3\left[(p_4/p_3)^{R/c_p} - 1\right] + T_2\left[(p_1/p_2)^{R/c_p} - 1\right]}{T_3 - T_2}. \tag{3.40}$$

Because $p_4/p_3 = p_1/p_2$ (Figure 3.16(b)), Equation (3.40) reduces to

$$\eta_{\text{th}} = 1 - \left(\frac{p_1}{p_2}\right)^{R/c_p}. \tag{3.41}$$

Typically the pump (or compressor, as it is more usually called) will produce a tenfold pressure increase, that is, $p_1/p_2 = 0.1$. For air, $R/c_p = 0.286$, and so η_{th} is 48.2 percent. In practice the efficiencies are lower than this because of internal friction and other losses.

Notice that to increase the efficiency of a gas turbine, the pressure ratio must be increased. This in turn will require higher and higher temperatures (Equation (3.35)). The highest temperature in the cycle is T_3; this is where the hot, high-pressure gas leaves the combustion chamber and is ready to be discharged through the turbine to do shaft work. Temperatures as high as 1,500 K can be produced here. The high-temperature limit is primarily fixed by the amount of thermal stress the turbine blades can tolerate before yielding. Much research is being carried out using advanced ceramic and composite materials for their construction. Once again design is dictated by thermodynamics: High efficiency requires high temperatures, just as was true for the case of our closed-cycle engine analysis (Chapter 2).

The gas turbine cycle of Figure 3.16 is often reconfigured, as shown in Figure 3.17(a). Here air enters the compressor, passes through the combustor and turbine, and is then exhausted into the atmosphere. However, the turbine shaft is connected to the compressor so that the compressor does not have to be supplied with a separate source of power; some of the work produced by the turbine is used to drive it. Equation (3.38) shows that the work produced by the turbine is proportional to $T_3 - T_4$, whereas that needed by the compressor is proportional to $T_2 - T_1$, a smaller temperature difference, because most of the temperature rise is in the combustor. Hence the work produced by the turbine is greater than

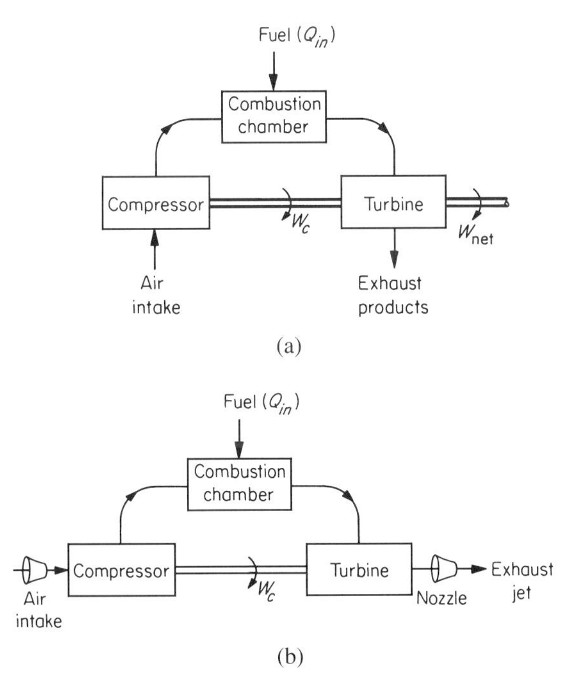

Figure 3.17 A gas turbine power generator (a) and a jet engine (b). They are similar, and jet engines can be modified to become megawatt power plants.

that needed by the compressor. Otherwise, of course, there would be no point in its design!

The gas turbine of Figure 3.17(a) can be further modified to make a jet engine, Figure 3.17(b). Here, the primary objective is to produce thrust to move the airplane, rather than to use the system as a power plant. Thus a high exit velocity is needed. Clearly for this use all the gas cannot be returned to the compressor via a heat exchanger; this is an open-loop system as opposed to the gas turbine power generator, which is usually a closed-loop system. The thrust, *Th*, of a jet engine is given by the relation

$$Th = \dot{m}(V_{\text{out}} - V_{\text{in}}), \tag{3.42}$$

where $\dot{m}$ is the mass flow of air through the engine. Although we have not derived this expression, its form is sensible; high thrusts will occur for large mass flow rates and large exit velocities (compared with entrance velocities). The mass flow rate, $\dot{m}$, is slightly higher for the gas leaving the jet than the gas entering, because fuel has been injected in the combustion chamber. However, the fuel-to-air ratio is quite low, around 5 percent, so in Equation (3.41) we have assumed $\dot{m}_{\text{in}} = \dot{m}_{\text{out}}$. The open-loop cycle is designed so that the gas leaves the turbine at relatively high pressure. Thus only part of the high pressure developed by the compressor is used to power the turbine; the rest is used to produce thrust. This high-pressure gas is accelerated (and expanded to atmosphere pressure) through a nozzle placed at the end of the engine. This results in very high exit velocities. The power produced by the turbine is used to power the compressor as well as to provide air conditioning, electricity, and so forth for the aircraft. Jet engines can produce tens of megawatts of power, enough for a sizable village. Although Equation (3.42) is quite straightforward, a full understanding of how the net thrust is produced requires a detailed analysis of the forces within the engine itself.

We have sketched here the basic principles of gas turbine power generation. Our analysis has been deliberately simple in order to keep these principles clear. For instance, we have assumed that c_p and c_v are constant and that the air at high temperatures and pressures obeys the perfect gas law. And we have assumed frictionless flow. In real design work, these assumptions must be lifted, and thus the efficiency for a real gas turbine will be lower than that calculated using Equation (3.41). Nevertheless, the principles will remain the same.

3.6 The Fluid Mechanics of the Internal Combustion Engine

So far in this chapter we have been dealing mainly with flow in pipes, be they straight sections with nothing but fluids inside or sections with turbines and compressors. We have shown that there are always energy losses when a fluid flows in a pipe. Enhanced turbulence around turbine and pump blades will greatly increase these energy losses, because further random velocity gradients will occur, producing higher shear stresses. Thus a very important part of turbine and compressor (pump) design is devoted to reducing the turbulence around the blades. The objective is to make $\dot{W}_f$ (Equation (3.33)) as small as possible.

What about the fluid mechanics inside the piston-cylinder engine that we analyzed from a thermodynamic viewpoint in Chapter 2? There are billions of these in operation, and an increase in efficiency of just a little, because of improvements in our understanding of the fluid mechanics, would make a great difference to world energy reserves.

In our discussion of turbulence (Section 3.2.2) we explained that the complicated turbulent motion produced random velocity gradients and hence random stresses inside the fluid. These stresses result in enhanced momentum transfer. Along with enhanced momentum transfer, turbulence also enhances heat and mass transfer. We automatically stir a fluid rapidly if we wish to mix its contents, be it cream in coffee or dye in a large vat. In order to promote quick and efficient combustion, we also require turbulence: The fuel (vaporized gasoline) and oxidant (air) must be thoroughly mixed so that as many molecules of air come into contact with as many molecules of fuel as possible. This occurs in the carburetor of a spark ignition IC engine and in the cylinder itself of a Diesel engine. Here, for simplicity we will confine our discussion to the latter type of engine. If all the fuel was on one side of the cylinder and all the air on the other, combustion would occur only over a relatively small interface, resulting in much less efficiency (Figure 3.18).

Efficient turbulent mixing in an engine cylinder is promoted by creating a swirling jet of fuel and air at the cylinder intake. This impinges on the walls of the cylinder (the no-slip condition holds here too) and the piston (which is moving out) to produce large-scale circulating patterns in the cylinder such as

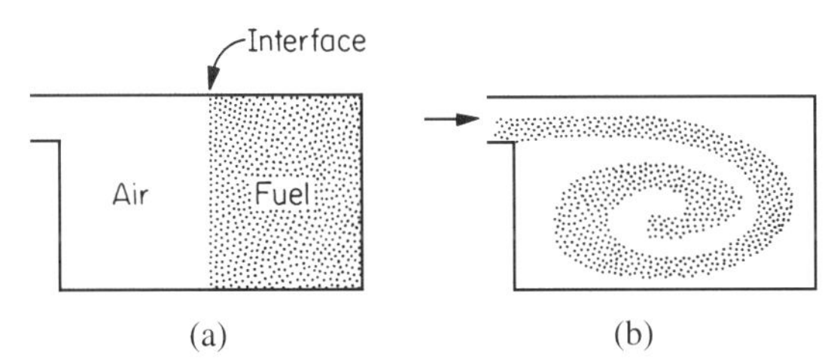

Figure 3.18 Inefficient (a) and efficient (b) mixing of fuel and air in a cylinder. Vortex structures with high turbulence levels enhance the mixing. Here the shaded area is the fuel. The area of the interface between the air and the fuel increases as the flow evolves in (b). Because combustion occurs at the fuel–air interface, (b) is much more efficient than (a), which has a smaller interfacial area. (Similarly, the interfacial area of the turbulent flow of Figure 3.2(b) is much greater than that of the laminar flow of Figure 3.2(a).)

shown in Figure 3.18(b). Such a flow is called a recirculating turbulent flow. We can crudely calculate the mixing time of these large turbulent eddies as follows. The fuel must be injected at least as quickly as the outward motion of the piston so that the combustion can take place in a small fraction of the cycle, at the beginning of the expansion phase (Figure 2.13). If we assume the piston is moving at 2,000 rpm and a characteristic length of the cylinder is 10 cm, then the average speed of the piston is $(2 \times 0.1)(2{,}000/60) = 6.7$ m/s (the factor of 2 appears because the piston moves back and forth during one cycle). Because the fuel must be injected at least as fast as the piston moves out, we will assume the speed, w, of the intake jet is of the same order, say 7 m/s. This jet swirls around the cylinder, forming a large-scale rotating flow over the dimension, ℓ, of a few centimeters. Since the swirling turbulent flow has a characteristic speed, w, and a characteristic length, ℓ, we can, from the dimensions of w and ℓ, determine the **turbulence time scale**, τ, of these largest eddies. It is

$$\tau \sim \ell/w. \tag{3.43}$$

For our example $\tau \sim 0.1/7 \sim 14$ ms. Here we have taken ℓ as 10 cm, the full length of the cylinder, although in practice it is somewhat shorter because the combustion occurs in the early part of the stroke. These calculations are only to determine the order of magnitude. The $\sim$ sign is short for "approximately equal to."

Equation (3.43) is very important. It says that if a turbulent eddy has a size ℓ and a characteristic velocity w, then it has a time scale of approximately ℓ/w. It must be emphasized that w is the motion of the largest eddies within the flow, not the mean velocity, V, of the flow as a whole. In order to further understand this point, consider a cumulus cloud (Figure 3.19). It may be moving along at 20 m/s(V), and it may have large-scale internal motions of, say, 1 m/s(w). If the cloud has a characteristic length (e.g., its depth or height) of, say, 100 m, then the time scale of the turbulence from Equation (3.43) is $\tau \sim 100/1$ s, which is nearly 2 min. If you watch a large cloud, you will see that it takes a minute or so (rather than the seconds that would result if V rather than w were used) for major changes to occur in its shape. This is its characteristic time scale of the largest eddies. For our piston engine example there is in fact no mean velocity because the flow is confined to a chamber, so $V = 0$. Here we have assumed that the characteristic speed of the large-scale turbulence, w, is approximately the same as the speed of the piston.

If we examine the structure of turbulence, we notice that there are not only large eddies but a whole range or spectrum of sizes. The time scale of the smaller eddies is smaller than that of the large ones. You will notice this again from watching clouds or waterfalls. You will observe that the small scales change more quickly than do the large scales. Dynamical theories of turbulence suggest

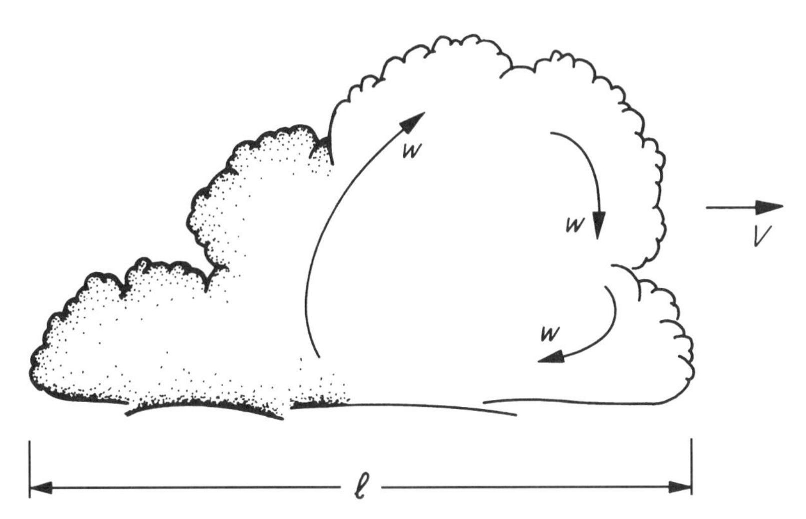

Figure 3.19 A cloud. The speed of the cloud as a whole is V, whereas the characteristic large eddy motion speed is w. The time scale of the large-scale motion (that which you see from the ground) is ℓ/w.

the large eddies break up into smaller and smaller eddies, until viscosity damps the motion of the very smallest eddies. This *cascade* of motion from large scales should take longer than τ, but in fact it does not take much longer because the smaller scales evolve faster (Figure 3.20). Thus τ, the turbulence time scale, may be interpreted at the *characteristic lifetime* of an eddy, or its *mixing time*.

The lifetime of a turbulent eddy is short; once the eddy has turned over, it has decayed. New eddies form in its place, but the identity of the original eddy is lost. While the eddy is turning over it passes all of its kinetic energy down the cascade to the very smallest eddies, which *dissipate* the turbulence energy, converting into internal energy (heat). In this respect turbulence is like a very rusty pendulum that comes to rest after only one period of oscillation, converting all of its kinetic energy into internal energy. Like the rusty pendulum, turbulence is an extremely efficient converter of kinetic energy into internal energy, and this is why it causes very large energy and pressure losses in pipes and machinery.

Our description of the flow inside an engine has been idealized in order to illustrate fundamental principles. For example, in real engines turbulence is sometimes deliberately suppressed in some regions and promoted in others. At present there is much research in calculating precisely the motion of the fluids inside an IC engine, as well as in the intake and exit valves, so that mixing rates and hence engine efficiency can be improved. Some of the largest computers are involved in these calculations, and we will conclude this chapter discussing the relation between computers and experiments in engine design, be it a turbine or an IC engine.

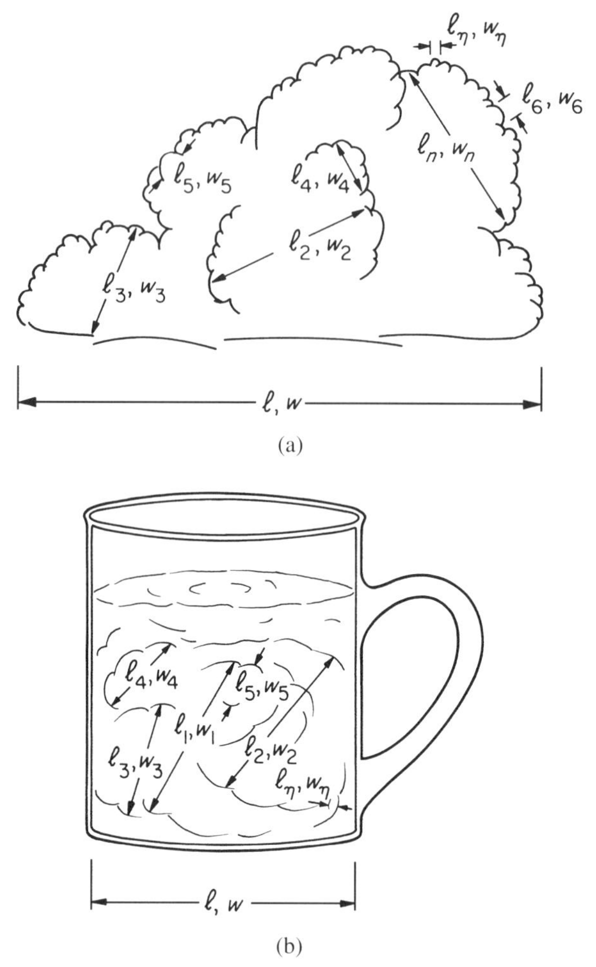

Figure 3.20 The turbulence cascade. Shown are a cloud (a) and a cup of rapidly stirred water (b). ℓ is the largest scale (the size of the cloud or cup), and successively smaller scales ($\ell_1 \ldots \ell_\eta$) are randomly embedded in the turbulent field. η (denoted as ℓ_η in this sketch) is the smallest scale (see Section 3.7). Notice that each length scale has a characteristic velocity and hence characteristic time scale. However, the overall energetics of the turbulence is determined by the largest length and velocity scales, ℓ and w respectively. The great British meteorologist L.F. Richardson wrote the following rhyme. "Big whirls have little whirls, which feed on their velocity, and little whirls have lesser whirls, and so on unto viscosity." The effect of viscosity on the turbulence is discussed in the next section.

3.7 The Use of Computers in Studying the Fluid Dynamics of Engines

The complex recirculating flow in a piston-cylinder, or the flow around the blades of a turbine or compressor, is very difficult to study. Placing sensors to measure the random fluid velocities and pressures in these systems is difficult. In recent years nonintrusive measurements have been done using lasers that can detect temperature, density, and velocity fluctuations. For this purpose the apparatus has to be specially constructed; for example, it must have transparent ports for the laser light, and the flow conditions are usually limited by the ability of the lasers to track them. These experiments are extremely costly and are generally carried out only by very large companies.

An attractive alternative to experiments is the use of large computers. Using the equations of fluid motion (which are described in the appendix), the full flow field inside an IC engine can be simulated. The task is enormous here too. In order to properly describe the real motion in an IC engine, all the scales of turbulence must be known. We know that the largest turbulent scales are of the order of 10 cm. What about the smallest scales? In the previous section the notion of a turbulence cascade was introduced; the large eddies break up into smaller and smaller eddies. These small scales are determined by the viscosity of the fluid itself, because if the Reynolds number of the eddies themselves becomes very small (less than 1), the eddies smear out or diffuse, losing their identity. However, if the energy of the large scales is made larger by increased stirring or mixing, the cascade penetrates to smaller and smaller scales. You will notice this if you put a little dye in a bath tub and stir. The more rapidly you stir, the more energetic the turbulence is, and the smaller is the size of the smallest eddies. This suggests that the size of the smallest eddies is not only determined by the viscosity, ν, but by the rate of energy input (per unit mass) to the large-scale turbulent motion. This energy input rate is just the energy of the large eddies divided by their characteristic time, which is w^2/τ. In fact the energy per unit mass of the large eddies is $(1/2)w^2$, but we are not interested in factors of two in this approximate analysis. Thus the smallest scales, which we will denote by the letter η, are a function of the viscosity, ν, and the rate at which turbulent energy is generated, w^2/τ. This can be more formally written as

$$\eta = \text{fn}(\nu, w^2/\tau).$$

Because $\tau \sim \ell/w$ (Equation 3.43), we can write

$$\eta = \text{fn}(\nu, w^3/\ell). \tag{3.44}$$

In spite of much research the precise dependence of w, ℓ, and ν on η has not yet been determined, but we can at least determine its order of magnitude by recognizing that both the left- and the right-hand side of Equation (3.44) must have the dimensions of length. The only way of forming a length from $\nu(\text{m}^2/\text{s})$ and w^3/ℓ (m^2/s^3) is to group them as follows:

$$\eta = \left[\frac{\nu^3}{\epsilon}\right]^{1/4}, \tag{3.45}$$

where I have used the symbol ϵ for w^3/ℓ. (See Problem 3.13 for a more familiar example of this technique, called *dimensional analysis.*) The length η, the smallest turbulence scale, is called the Kolmogorov length, after the great Russian mathematician and scientist who first defined it in 1941. Note that Equation (3.45) does not tell us the precise value of the smallest scale because the dimensional analysis does not preclude a constant of proportionality (see Problem 3.13). However, experiments show it does give the right order of magnitude, and this is enough for most calculations.

Returning to our piston-cylinder example, and using the values of 7 m/s for w, 10 cm for ℓ, and 10^{-4} m^2/s for ν (a slightly higher value than previously used because here the gas is very hot and ν increases with temperature for gases), we find from Equation (3.45) that $\eta \sim 1.3 \times 10^{-4}$ m, or approximately 0.1 mm. This is small indeed! Our calculation suggests that if you look closely at a photo of the motion of gases in an IC engine, the complex grainy turbulent structure will cease for scales below about 0.1 mm. At scales smaller than this, the motion will look smooth. Although 0.1 mm is small, it is still very large compared with atomic or molecular dimensions. Although turbulence causes complex wrinkled motion at many scales, its motion does not penetrate down to molecular dimensions.

Our calculations show, then, that the volume of the largest eddy in a cylinder is approximately $(10\text{ cm})^3$, or approximately 10^{-3} m^3, and the volume of the smallest eddy or disturbance is $(0.1\text{ mm})^3$, or approximately 10^{-12} m^3. I have deliberately rounded off numbers; here we are working to determine order-of-magnitude values only. This suggests that to fully describe the motion inside a piston-cylinder, our computer must store a velocity vector for $10^{-3}/10^{-12} = 10^9$ positions in the cylinder. This is known as the number of grid points. A photo of gas flow inside a cylinder is compared with a computer simulation in Figure 3.21. In order to follow the evolution of the flow the velocity, pressure, and density of the flow at a particular instant are specified for all the grid points. The equations of motion are then *time-stepped.* In fractions of a millisecond the vector picture of the fluid motion will change, and so the velocity field will have to be recalculated again and again. I have not written down the equations of motion; they are quite complicated, involving terms that describe the viscosity, pressure, and acceleration within the fluid. Their form is discussed in the appendix.

We see that what goes on inside a piston-cylinder is really quite extraordinary; the fluid motions are just as complex as those of the weather and galaxies. We have shown that to generate this motion in a computer requires a billion grid points where the velocity vector must be defined. In fact a more detailed analysis shows that this number can be considerably reduced by making approximations concerning the smallest scales of motion. Even so, only the very largest computers are capable of these calculations, which take many hours to simulate what happens in an engine in a few milliseconds. Engineers work with computer

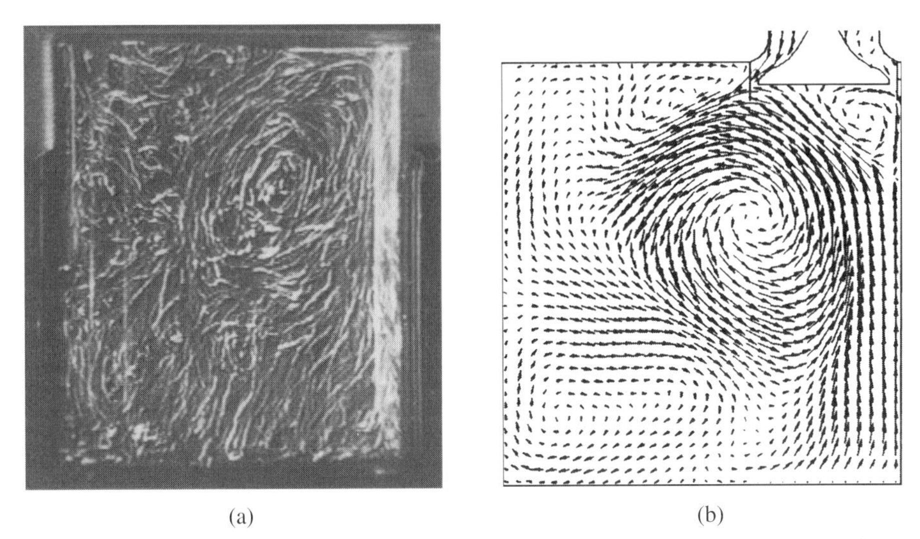

(a) (b)

Figure 3.21 (a) A photo of flow inside a piston-cylinder. (b) A computer simulation of the same flow. The beginning of each tiny vector is roughly equivalent to a grid point. Not all of the fine details of the flow are shown because this would require too many grid points. (Parts (a) and (b) courtesy of Dr. D. Haworth, General Motors, and reprinted with permission from SAE paper No. 941871 © 1994 Society for Automotive Engineers, Inc.)

codes, and after describing the motion for a conventional engine, they do small modifications, changing, for example, the amount of swirl of the incoming flow. They then recalculate the flow inside the cylinder to see whether turbulence levels have changed and, if so, whether the mixing has been enhanced. Similar and equally large and costly calculations are done for the flow around turbines, compressors, and combustors in gas turbine engines.

3.8 Summary and Discussion

Internal friction is an inherent characteristic of fluid flow. At low Reynolds numbers when the flow is laminar, these frictional effects are less important than when the flow becomes turbulent at high Reynolds numbers. Much of the fluid flow of interest to mechanical engineers is in fact at high Reynolds numbers, such as in engines and in the atmosphere. (One important exception is in lubricating bearings, where the Reynolds number is low because of the small dimensions and high viscosity of oil. Here the flow is laminar.) When turbulence occurs, random stresses occur due to the random nature of the turbulent eddies. These eddies increase the fluid friction and thus the pressure drop along pipelines. They also reduce the efficiency of turbines and pumps. On the other hand, turbulence also results in increased mixing rates, and this is beneficial because, for example, in an internal combustion engine or in the combustor of a jet engine, rapid mixing is needed in order to have efficient combustion. Because of these two faces of turbulence, design is sometimes compromised, with mixing and drag being played off against each other.

I have attempted in this chapter to introduce you to those aspects of fluid dynamics relevant to engine design. We will see in Chapters 4 and 5 that they will be used again to describe turbulent heat transfer and pollution in the atmosphere. There are, however, many other very important aspects of fluid motion that we have not discussed here, such as the fluid dynamics of flight and the motion of fluids at very high velocities when shock waves occur. These are not part of our story; they will be studied in later courses.

You may have noticed that when we dealt with thermodynamics both in Chapter 2 and again in Sections 3.4 and 3.5 of this chapter, rather precise calculations were done. For example, efficiencies were determined to one or two significant figures. This is because we were idealizing, assuming that flows were frictionless and so on. Once we started to examine the reality of the fluid motion, calculations became more approximate and often were done only to an order of magnitude. As I explained in the introduction, this is typical of the nature of engineering calculations, and no matter how sophisticated analytical techniques become, order-of-magnitude analysis will always play a major role in frontier design. Only when these calculations are done can the detailed design begin.

3.9 Problems

3.1 The Reynolds number is the product of a characteristic velocity and length scale, divided by the kinematic viscosity (which is a property of the fluid and is fixed). Calculate the Reynolds number for the following:
- (a) a professionally pitched baseball,
- (b) a shark swimming at full speed,
- (c) a passenger jet liner at cruise speed,
- (d) the wind flow around Mount Mauna Kea in Hawaii,
- (e) natural gas (same viscosity and density as air) flowing in a 2-m-diameter pipeline at 40 kg/s,
- (f) a very small bug flying in air.

Make reasonable estimates of characteristic speeds and lengths, and state them. State whether the flow will be turbulent. Assume that ν for air and water are respectively 10^{-5} m^2/s and 10^{-6} m^2/s in all problems.

3.2 When a uniform flowing fluid passes over a surface, the no-slip condition must hold (Figure 3.4(b)). The flow at the wall stops, but the fluid far away keeps going at its undisturbed speed. There results a region with a pronounced velocity gradient. This is called a **boundary layer**. As the flow evolves downstream, the boundary layer will thicken, because momentum

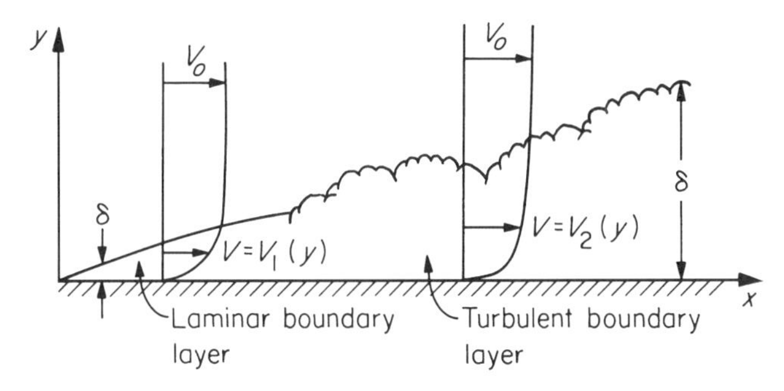

Figure 3.22 Figure for Problem 3.2.

diffuses toward the wall (Figure 3.22). Initially the boundary layer will be thin and the flow will be laminar, but as it thickens it will undergo transition to turbulence. For this flow there are two length scales, the thickness of the boundary layer, δ, and the distance along the plate, x. The relation between δ and x is given by

$$\frac{\delta}{x} = \frac{5.48}{(Re_x)^{1/2}},$$

where $Re_x \equiv xV_0/\nu$. This applies to the laminar part of the flow only and breaks down when $Re_\delta \equiv \delta V_0/\nu \sim 2{,}000$. For an air flow of 4 m/s coming into contact with a flat tabletop, what will be the depth of the boundary layer when the flow becomes turbulent? How far along the surface will the flow become turbulent? Redo the calculation for air going over an aircraft wing at 250 m/s (approximately 500 mph). Assume the aircraft wing is approximately a flat plate for this example. The viscosity of air is 10^{-5} m^2/s. From this you will see that boundary layers are very thin and become even more so as the Reynolds number increases.

3.3 Our discussion of pipe flow showed that transition from laminar to turbulent flow depends only on the Reynolds number and not on the actual value of the diameter, speed, or viscosity. When two flows of the same geometry have the same Reynolds number, they are *dynamically similar*. Thus, if you take a photo of turbulent flow of, say, air in a large (glass) pipe with a slow velocity and another of water in a small pipe at another velocity, they will look the same if their Reynolds numbers are the same. However, if one has a very high Reynolds number, say 10^6, and the other is somewhat lower, say 10^4, although both will be turbulent their structure will look somewhat different. The higher Reynolds number flow will look more turbulent: It will appear to have more details in its random structure. When engineers wish to design structures such as turbines or airplanes or buildings, in order to determine forces and mixing rates their model must have the same Reynolds number as the full scale. Consider the following problem.

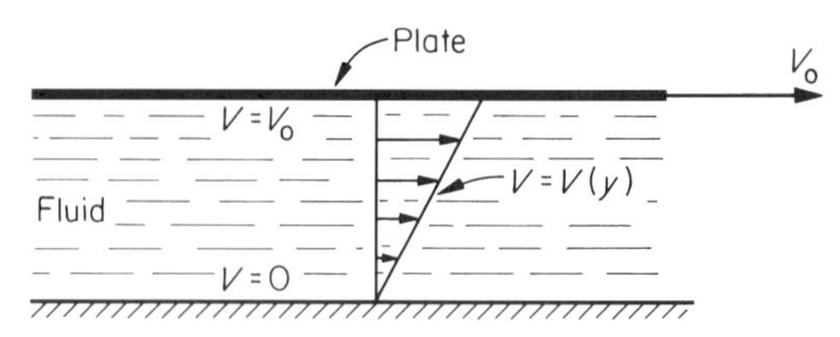

Figure 3.23 Figure for Problem 3.4.

An engineer wants to study the air flow around a large building that is in the form of a cube. Its volume is one million m^3. Clearly the building is too large to place in a wind tunnel! Instead the engineer decides to test a cubic model in a water tunnel. She wants to have the same Reynolds number for the model as for the actual building. Taking the expected wind speed for the building to be 5 m/s, ν for air to be 10^{-5} m^2/s, ν for water to be 10^{-6} m^2/s, and the model to have a volume of 8 m^3, what should the water speed past the model be?

3.4 If a flat plate sitting on the top of a fluid is pulled at speed V_0 parallel to a stationary plate, and the Reynolds number is low, the resulting flow is called laminar Couette flow (Figure 3.23). The no-slip condition must hold for both plates. Therefore, the fluid speed adjacent to the top plate must be V_0; that adjacent to the bottom plate, zero. A linear velocity profile is observed, and thus Equation (3.6) shows that the stress must be constant throughout the fluid. All the stress needed to move the top plate is imparted to the plate below.
Consider an oil of viscosity 10^{-4} m^2/s inside a bearing of clearance 0.3 mm. What force must be applied to the upper surface (which has an area of 0.5 m) to keep the plate moving at 2 m/s? How much power is required? What is the Reynolds number of this flow? State the criterion you used for selecting the characteristic velocity. Assume the density of the oil is 700 kg/m^3.

3.5 Show that for laminar flow in a pipe, the average velocity is one-half the center-line velocity. The volume flow rate of fluid flow is given by the integral $\int V dA$, where dA is the area of a differential annulus, that is, $dA = 2\pi r dr$. Show that for laminar flow in a pipe, the volume flow rate is $-(\pi R^4/8\mu)(dp/dx)$. What are its dimensions?

3.6 Engineers plan to build a pipeline 1,000 km in length to transport water to a drought-stricken region. They wish to have a volume flow rate of 10 million gallons (3.79×10^4 m^3) per day, and the maximum pipe diameter they can use is 0.5 m. The engineers do not wish to place a pump more frequently than every 20 km.
Determine the average velocity of the water in the pipe and the Reynolds number. Is the flow laminar or turbulent? Then, using Equation (3.15),

determine the pressure gradient along the pipe and the total pressure drop over its full length. Finally, determine the power of the pumps needed to keep the pressure up. The density and kinematic viscosity for water are 1,000 kg/m^3 and 10^{-6} m^2/s respectively.

3.7 A supply pipe for a gas turbine increases from 2 cm to 6 cm in diameter. Because of heating, the air density changes from 1.2 kg/m^3 to 0.6 kg/m^3. The average inlet velocity is 10 m/s. An engineer measures the average outlet velocity to be 1.8 m/s. She suspects a leak. Is she right? Explain your answer.

3.8 Using the ideal gas law and the relationships between specific heats for a perfect gas ($du = c_v dT$ and $c_p - c_v = R$), show that for an ideal gas with no elevation change, the Bernoulli equation (Equation (3.21)) becomes

$$c_p T + \frac{1}{2} V^2 = \text{constant}.$$

Consider air being pushed through a small hole from a syringe. The speed of the air coming from the hole is 60 m/s, while its speed in the syringe itself can be considered to be zero since its diameter is much larger than that of the hole. If the air temperature in the syringe is 300 K, what is the temperature of the air coming from the hole? Why is it lower? c_p for air is 10^3 J/(kg K).

3.9 Figure 3.24 shows a 1-cm diameter water siphon. Use the Bernoulli equation to determine the flow rate of the water. Assume that at the discharge the fluid pressure is atmospheric (10^5 Pa) and $\rho = 1{,}000$ kg/m^3. If the discharge level and water level were held constant, but the tube was lengthened so that the height of the horizontal arm was increased, would the flow rate be affected?

3.10 Consider an air turbine designed to produce 1 MW of power. The maximum pressure allowed is 30×10^5 Pa (30 atmospheres). Assuming c_p is constant, $\Delta h = c_p \Delta T$. (See the discussion following Equation (3.34).)

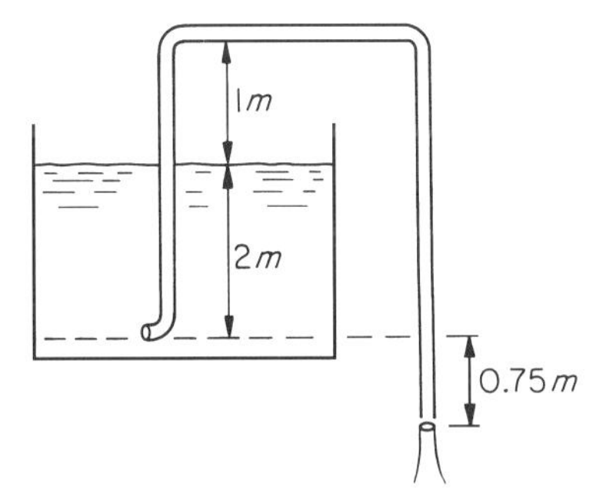

Figure 3.24 Figure for Problem 3.9.

Thus, the work of a turbine can then be written as

$$\dot{W}_s = c_p(T_2 - T_1)\dot{m}.$$

What is the flow rate and the average inlet velocity? The pipe diameter is 1 m, and the discharge is to atmospheric pressure and temperature. Assume that the perfect gas law holds and the fluid is air ($c_p = 1{,}000$ J/(kg K); $R = 287$ J/(kg K)).

3.11 The p-v diagram (Figure 3.16) shows an ideal Brayton cycle that is used for power generation. Consider a specific example for which the air enters the compressor at 1 atmosphere (10^5 N/m^2) and 20°C. The pressure ratio (p_2/p_1) is 6:1, and the air leaves the combustion chamber (and therefore enters the turbine) at 1,000 K (position 3 in Figure 3.16(b)). Assuming that the air behaves as a perfect gas and c_p and c_v are constant (1.0×10^3 J/(kg K) and 0.7×10^3 J/(kg K) respectively), determine the following:
(a) The pressure and temperature at all points of the cycle.
(b) The compressor work input (1–2 in Figure 3.16).
(c) The turbine work output (3–4 in Figure 3.16).
(d) The efficiency of the cycle.

3.12 A massive volcano produces a plume of order 3 km. Satellite remote velocity sensing instruments determine the large-scale eddy motion to have a velocity of 2 m/s. Assuming the density of the plume is 2 kg/m^3 (it is mostly gaseous), determine the kinetic energy of the turbulence. What is the characteristic time scale of the turbulence? What is the power of the turbulence? What happens to it?

3.13 In order to show that the concept of matching the units enables us to determine the relationship between a number of variables (as was done in going from Equation (3.44) to (3.45)), consider the problem of determining the period of a frictionless pendulum. Suppose we have forgotten Newton's laws, but we assume that the period, τ, must be a function of the length of the pendulum, ℓ, and g, the gravitational acceleration, that is,

$$\tau = \mathrm{fn}(\ell, g).$$

Because the right-hand side of this equation must have the unit seconds, determine its form. What is the value of the constant by which this answer differs from the exact solution determined from Newton's laws? (Note: if we had assumed $\tau = \mathrm{fn}(m, \ell, g)$ we could not have found dimensional consistency between the left- and right-hand side. Clearly the period is not a function of the mass.)

3.14 For the volcanic plume of Problem 3.12 determine the smallest scale of the turbulent motion. Determine the Reynolds number of the motion (Re_ℓ)

based on the velocity of the eddy motion and the plume scale. Assume $\nu = 10^{-5}$ m^2/s. Do the same calculations for the cumulus cloud described in Section 3.6 ($w = 1$ m/s, $\ell = 100$ m). Assume here too that $\nu = 10^{-5}$ m^2/s. Finally, do the same calculation for stirring a cup of water. Here $\nu = 10^{-6}$ m^2/s and you may make your own estimates of w and ℓ. Does there appear to be any relationship between the ratio ℓ/η and Re_ℓ? (Plot the 3 data points in log-log coordinates.) Interpret your result in terms of the way we picture turbulence.

Symbols

A	area	m^2
c_p	specific heat at a constant pressure	J/(kg K)
c_v	specific heat at a constant volume	J/(kg K)
d	diameter	m
E	total energy (kinetic + potential + internal)	J
e	total energy per unit mass	J/kg
g	acceleration due to gravity	m/s^2
L	pipe length	m
ℓ	turbulence length scale	m
m	mass	kg
$\dot{m}$	mass flow rate	kg/s
p	pressure	Pa
Q	heat interaction	J
$\dot{Q}$	rate of heat transfer (or interaction)	W
q	heat interaction per unit mass	J/kg
$\dot{q}$	rate of heat transfer (interaction) per unit mass	W/kg
R	pipe radius	m
R	gas constant (section 3.5)	J/(kg K)
Re	Reynolds number	
r	radial direction (in a pipe)	m
T	temperature	K
Th	thrust	N
t	time	s
U	internal energy	J
u	internal energy per unit mass	J/kg
V	velocity	m/s
V_{ave}	average fluid velocity (over pipe diameter)	m/s
$\overline{V}$	velocity averaged over a period of time (mean velocity)	m/s
v	specific volume	m^3/kg

W	work interaction	J
$\dot{W}$	rate of work transfer	W
W_f	friction work	J
W_s	shaft work	J
w	turbulence velocity scale	m/s
w_f	friction work per unit mass	J/kg
w_s	shaft work per unit mass	J/kg
$\dot{w}$	rate of work transfer (interaction) per unit mass	W/kg
x	streamwise direction	m
y	direction normal to the stream	m
z	height	m
δ	boundary layer thickness	m
ϵ	turbulence energy dissipation rate per unit mass	W/kg
η	Kolmogorov length	m
η_{th}	thermal efficiency	
μ	dynamic viscosity	kg/(m s)
ν	kinematic viscosity	m^2/s
ρ	density	kg/m^3
τ	turbulence time scale	s
τ_{sh}	shear stress	N/m^2

4

Heat Transfer

In the previous chapter we saw the importance of fluid dynamics in the analysis and design of engines. In the next chapter we will see its importance in understanding the motion of the atmosphere. Both in the engine and in the atmosphere the motion is affected by temperature differences that cause heat to be transferred from one region to another. In Chapter 2, on thermodynamics, we were interested in the magnitude of the heat transfer, but we were not particularly concerned with the mechanics of how it occurred or the rate at which it occurred. Clearly these are important issues because if the heat transfer can be improved (by being made closer to a reversible process and by minimizing losses to the environment) the design of our engines will be better. In this chapter we will study the way the heat transfer process occurs.

Consider first the piston-cylinder engine whose thermodynamics occupied so much of our attention in Chapter 2 and whose internal fluid motion we discussed at the end of Chapter 3. As in Chapter 2 we will assume that the heat addition is by means of an external hot plate, rather than by the more complex process of the ignition of a fuel and air mixture. When the hot plate is placed at the bottom of the cylinder (Figure 4.1), heat is transferred through the metal cylinder head by means of **conduction**. The internal energy of the atoms in contact with the hot plate increases, and this energy is then transferred (without any migration of the atoms themselves) to adjacent layers until the temperature of the whole cylinder head is increased. In turn a layer of air very close to the cylinder head is also heated by conduction. In a sense the atoms in each layer rub shoulders with each other, passing their increased energy from layer to layer. This, you will recall, is similar to the way momentum was transferred in laminar flow by means of the viscous stress. Inside the cylinder, however, the difference in temperature between the gas close to the hot plate and the cool gas higher up in the cylinder will cause large-scale motion of the gas. This motion (which is usually turbulent) will allow for the higher-temperature fluid to be transported to regions of cooler

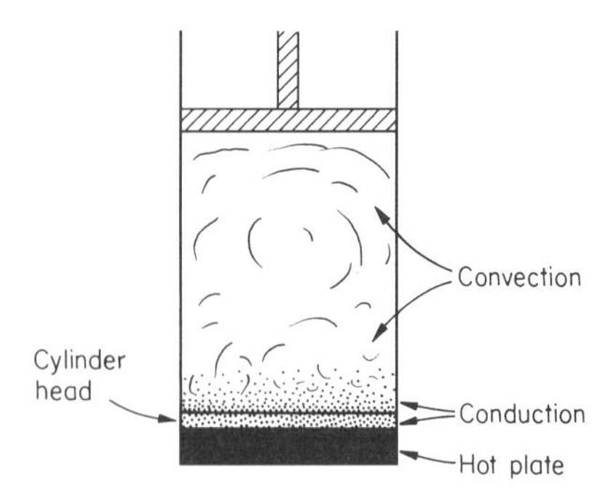

Figure 4.1 Conductive and convective heating of a gas inside a cylinder.

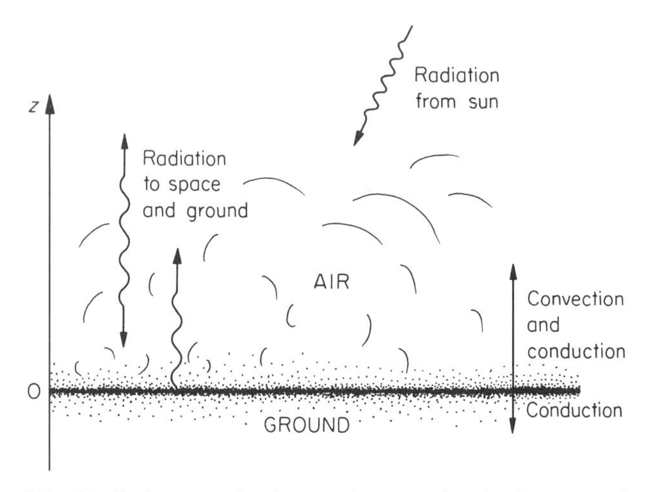

Figure 4.2 Radiation, conduction, and convection in the atmosphere.

temperature, thereby mixing the gas inside by means of **convection**. Convective heat transfer implies that the medium is in motion. Thus, convection can occur only in fluids (be they liquids or gases). On the other hand, conduction can occur in any medium, be it solid or fluid.

Consider now the atmosphere (Figure 4.2). The sun's light heats the atmosphere and the ground. The mode of heat transfer from the sun is called **radiation**. When the radiant energy reaches the ground, conduction occurs through it and also through the air above. At the same time convection transports the heat to higher regions. Like the sun, the atmosphere and the ground also radiate, so the heat transfer interactions are quite complicated. Understanding heat transfer in the atmosphere is important, not because we can alter it to make it better (like an engine) but because we *are* altering it by means of pollution. We are still unsure whether the way we are altering it will really matter in the future, but unless we really understand the mechanisms, it is possible that irreparable damage will have been done before we have decided what legislation to implement to improve the situation. Heat transfer, for both the engine and the atmosphere, is a very important subject.

It is apparent from this discussion that heat transfer between bodies of unequal temperature occurs by three mechanisms: conduction, convection, and radiation. Often these occur together. Conduction and convection require a material medium (gases, liquids, or solids for conduction and gases or liquids for convection), and their mechanisms are linked. Radiative heat transfer can occur through a vacuum as well as through a medium. We will examine each in turn. In Chapter 5 we will link them all when we study the greenhouse effect.

4.1 Conduction

In order to develop the equations for heat conduction, we begin by considering a simple example: a slab of material of area A and width x, heated on one side (Figure 4.3). There will be heat transfer in the direction shown. The heat transfer rate will be proportional to the temperature difference $T_2 - T_1 = \Delta T$ and the area of the slab: Increasing ΔT will increase the heat transfer rate, and increasing A will allow a larger heat transfer for a given ΔT. If the slab is very thin, the heat transfer rate will be large, but if the slab is thick, it will be smaller. Thus, the heat transfer rate will be inversely proportional to the slab width, x. Hence we can write

$$\dot{Q} \propto \Delta T \frac{A}{x} \qquad \text{or} \qquad \dot{Q} = k \Delta T \frac{A}{x}, \tag{4.1}$$

where $\dot{Q}$ is the rate of heat transfer (watts). The proportionality constant, k, is called the thermal conductivity, and its units are W/(m K).

We can write Equation (4.1) in a more general form by defining $\dot{q}_c \equiv \dot{Q}/A$. This is the rate of heat transfer per unit area. Note that the area A is defined such that the normal to A is in the same direction as the temperature difference. Thus

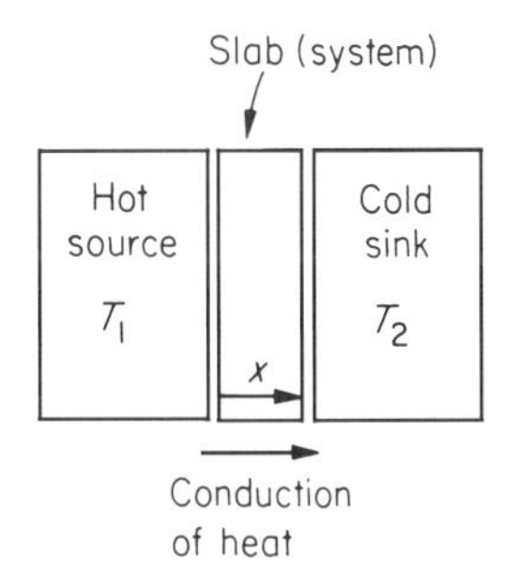

Figure 4.3 Heat conduction. Here we take the surface of the slab as the control surface.

in Figure (4.3) A is the area of the slab face whose normal is in the x direction. Further, we recognize that $\Delta T/x$ is a gradient of temperature, that is, it is the change of temperature with distance, and in the differential limit it is written as dT/dx. Thus Equation (4.1) becomes

$$\dot{q}_c = -k\frac{dT}{dx}. \tag{4.2}$$

Equation (4.2) is known as Fourier's law of heat conduction. The minus sign is required for $\dot{q}_c$ to be positive because the heat flow is from hot to cold; thus, dT/dx is negative (Figure 4.3).

Equation (4.2) is of the same form as Equation (3.6), which related the fluid stress to the velocity gradient. This is no coincidence. You will recall that the interpretation of Equation (3.6) was that the rate of momentum transport per unit area (the shear stress) is proportional to the momentum (or velocity) gradient and the proportionality constant is the dynamic viscosity, μ. Equation (4.2) shows that the rate of heat transport (or transfer) per unit area is proportional to the temperature gradient and the proportionality constant is the thermal conductivity, k. Just as momentum is transferred down a gradient (that is, from high to low values), so is heat. In fact the notion of heat transfer down a temperature gradient is much more intuitive to us than that of momentum transfer down a velocity gradient.

As was the case for Equation (3.6), Equation (4.2) holds for many but not all substances. Fortunately, for most common substances such as water, air, oil, steel, and concrete it is obeyed very well. Of course, unlike the case of the viscosity relationship (Equation (3.6)), Equation (4.2) holds for solids as well as for fluids. The thermal conductivity for steel is approximately 30 W/(m K) (it depends on the type of steel and its temperature), whereas k for water and air are 0.6 and 0.03 W/(m K) respectively (and here too there is a temperature and pressure dependence). Thus for a 100 K temperature difference across 1 cm, the heat transfer rate will be, from Equation (4.2), 3×10^5 W/m^2, 6×10^3 W/m^2 and 300 W/m^2 for steel, water, and air respectively. This result suggests why it is less damaging to put your hand in a hot oven than on a hot iron; for the same temperature difference the heat transfer rate is 1,000 times greater for the iron.

Equation (4.2) conveys the essence of the physics of heat conduction. However, in both physics and engineering we often wish to determine the temperature distribution in a solid or fluid and its rate of change with time. In order to do this we will return to the first law of thermodynamics as a rate equation. Consider the differential slab in Figure 4.4(a) as the control volume. The slab may be a solid, liquid, or gas. We will assume, for generality, that it is a gas (which expands on heating). Our result will then easily be modified for liquids and solids, which are (almost) incompressible. We are particularly interested in the rate of change of temperature with time, so this is not a steady-state problem. The heat entering on

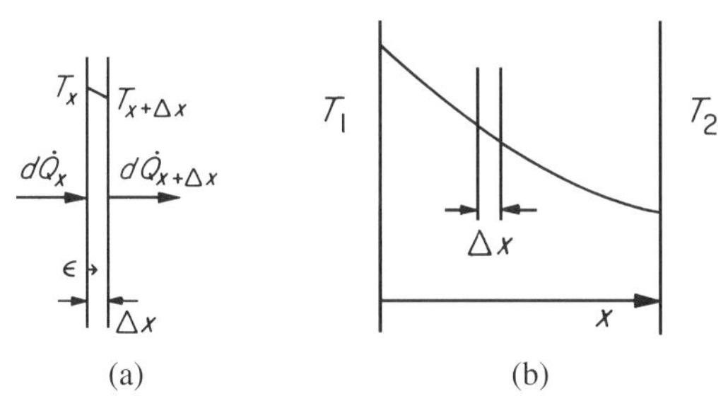

Figure 4.4 The use of differential slabs to develop the heat conduction equation.

the left-hand side, $d\dot{Q}_x$, may not in general equal that leaving on the right-hand side, $d\dot{Q}_{x+\Delta x}$*. (Consider what happens when a hot plate is initially applied to the left side of the slab in Figure 4.4(a).) Thus, in a general way we may write that the net rate of heat transfer into the control volume, $d\dot{Q}_{\text{in}}$, is

$$d\dot{Q}_{\text{in}} = d\dot{Q}_x - d\dot{Q}_{(x+\Delta x)}. \tag{4.3}$$

If nothing is changing with time, this term will of course be zero (unless there is some new heat source inside the slab such as an electric resistor). (See Problem 4.3.)

We now make an estimate of $d\dot{Q}_{\text{in}}$ in terms of the gradient of $\dot{Q}_x$ across the control volume. Here we use a general approach that you will use repeatedly in courses to come, so it is worthwhile carefully pondering it. Let us assume that $\dot{Q}_x$ is varying linearly across the distance Δx. Thus, if we move from the left face an extremely small distance, say ϵ, to the right, $\dot{Q}_x$ will increase by say, γ (Figure 4.4(a)). If we move 2ϵ, $\dot{Q}_x$ will increase by 2γ and so on. This means that the heat transfer gradient $d\dot{Q}_x/dx$ is just

$$d\dot{Q}_x/dx = \frac{d\dot{Q}_{x+\Delta x} - d\dot{Q}_x}{\Delta x},$$

and thus Equation (4.3) becomes

$$d\dot{Q}_{\text{in}} = -\frac{d\dot{Q}_x}{dx}\Delta x. \tag{4.4}$$

Now, you may say that this is not a very general approach and that $\dot{Q}_x$ will, in general, not vary linearly with x. For instance doubling ϵ may quadruple $\dot{Q}_x$. But consider this. We have taken Δx, the thickness of the slab, as arbitrarily small. Thus, a linear approximation over this small length should be adequate

* In order to be consistent with Chapter 2, I should use the notation $\delta\dot{Q}$ rather than $d\dot{Q}$ to signify that heat transfer is an interaction between the system and its surroundings. However, the d notation is more convenient for the development in this section.

even if $\dot{Q}_x$ does not vary linearly over an entire slab, as will generally be the case (Figure 4.4(b)). Equation (4.4) should hold better and better as $\Delta x \rightarrow 0$. In fact it can be shown that Equation (4.4) is precise for most heat conduction problems. The right-hand side of Equation (4.4) is known as the first term in a Taylor series expansion. In some problems, where there are sharp discontinuities in temperature, higher-order terms are needed, but this is rare.

Let us return to the first law of thermodynamics, which is, for the differential slab of Figure 4.4(a):

$$d\dot{Q}_{\text{in}} = \frac{dU + p\,d\forall}{dt}. \tag{4.5}$$

Here we have differentiated the first law with respect to time and have allowed for the differential slab to expand, doing some $pd\forall$ work on its surroundings as its temperature changes (Equation 2.29). Because our slab is at rest and is free to expand, it is reasonable to assume that the process occurs at a constant pressure (the surroundings being at the same pressure as the slab). Hence, Equation (4.5) becomes

$$d\dot{Q}_{\text{in}} = \frac{dH}{dt} = mc_p\frac{dT}{dt}. \tag{4.6}$$

Here $H (\equiv U + p\forall)$ is the enthalpy, and m is mass of the slab. Because p is constant, $dH = dU + pd\forall$. Recall also (p. 94) that $dH = mdh = mc_p dT$, where c_p is the specific heat at a constant pressure.

Substituting $-(d\dot{Q}_x/dx)\Delta x$ for $d\dot{Q}_{\text{in}}$ (Equation (4.4)) on the left-hand side of Equation (4.6), and using Equation (4.2) with $\dot{q}_c = \dot{Q}/A$, Equation (4.6) becomes

$$\frac{d}{dx}\left(k\frac{dT}{dx}\right)A\Delta x = \rho\Delta x A c_p\frac{dT}{dt}. \tag{4.7}$$

Here we have written the mass m as $\rho\Delta x A$, where A is the area of the differential slab. If k is assumed constant with temperature, a reasonable approximation in many problems, then Equation (4.7) becomes

$$k\frac{d^2T}{dx^2} = \rho c_p\frac{dT}{dt}. \tag{4.8}$$

Notice that in Equation (4.8), T varies with both x and t, and for this reason we really must use partial derivatives in order to recognize that T is a function of more than one variable. Equation (4.8) must be written in the form

$$k\frac{\partial^2T}{\partial x^2} = \rho c_p\frac{\partial T}{\partial t}. \tag{4.9}$$

If you have not studied partial derivatives yet, do not worry. Here you need only recognize that it is a warning that you must keep track of both the spatial and temporal variations in temperature. Equation (4.9) is often written as

$$\frac{\partial T}{\partial t} = \alpha \frac{\partial^2 T}{\partial x^2}, \tag{4.10}$$

where $\alpha \equiv k/(\rho c_p)$ is called the thermal diffusivity. Its units are m^2/s – the same as kinematic viscosity. In gases the values of ν and α are almost the same; in air their ratio ν/α (called the Prandtl number, *Pr*) is approximately 0.7. Here the momentum and heat transport mechanisms are similar. However, in liquids there is a very great variation in *Pr*. Furthermore, liquids (and solids) are almost incompressible. Therefore, for liquids and solids we can write: $c_p = c_v = c$ and $\alpha = k/(\rho c)$.

Equation (4.10) is one of the most important equations in all of classical physics and engineering. It is called the **heat conduction equation**. It states that the rate of change of temperature with time at any point is proportional to the comparative surplus or excess of temperature at that point. Thus, if there is a hot spot on the slab, its temperature will even out with time, under the condition, of course, that the medium can conduct heat ($\alpha \neq 0$).

Equation (4.10) not only holds for temperature, it also holds for concentration. Here it is called the diffusion equation and it is written in the form

$$\frac{\partial F}{\partial t} = D \frac{\partial^2 F}{\partial x^2}, \tag{4.11}$$

where F is the concentration of some constituent and D is its diffusivity. For example, if ink is dropped on a plotting pad, it will spread out, moving from the region of high to low concentration as shown in Figure 4.5. (Equation (4.11) will have to be extended to three dimensions, and we will do this in a moment.)

Equations (4.10) and (4.11) are unidirectional; they are irreversible. Heat or mass flows from high to low concentrations, not in the reverse direction. This is a second-law issue, and the reversibility is implicit in the Fourier heat conduction equation (Equation (4.2)). Its interpretation lies in the molecular nature of matter. If you consider a vessel of water (Figure 4.6) with a high concentration of dye on the left, the dye molecules will diffuse to the right as a result of their molecular motion. You should realize that each molecule acts independently; they do not have a conspiracy to move as a group. All are in random motion, each being knocked around by the water molecules, which are also in random motion. They have no preferred direction as individuals, but because their concentration is high on the left, chance random motion to the right will result, after a long period, in a migration of dye concentration as a whole. To convince yourself of this, assume for simplicity that there is equal probability of a molecule going up/down or left/right and trace the path of a number of these acting independently. (Choose a group of dye molecules that lie to the edge of where they are concentrated.)

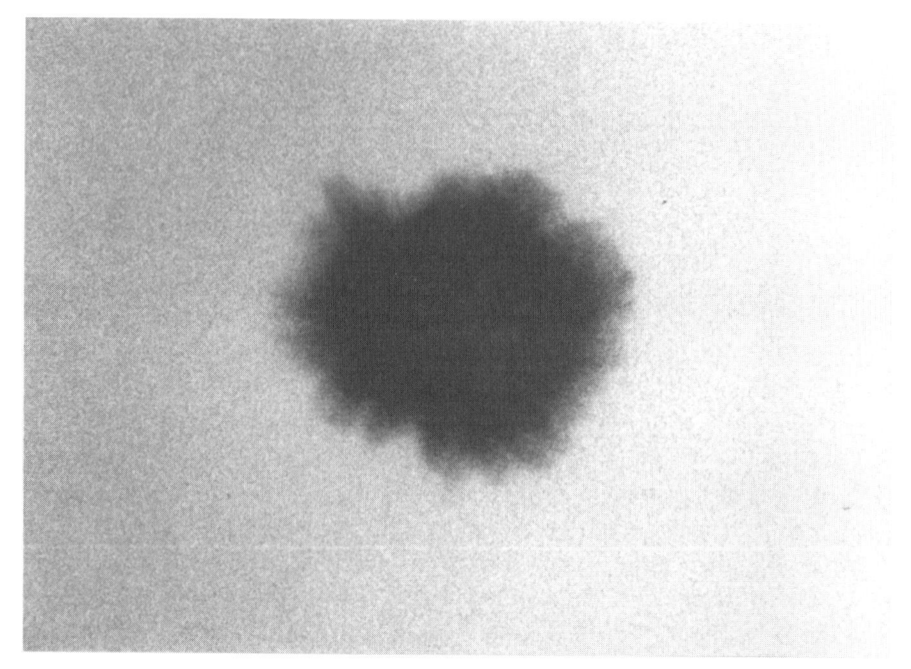

Figure 4.5 Ink diffusing on blotting paper from the region of highest concentration to regions of no ink. The equation of ink diffusion is precisely the same form as that of heat conduction. (Courtesy of Mr. E.P. Jordan, Cornell University.)

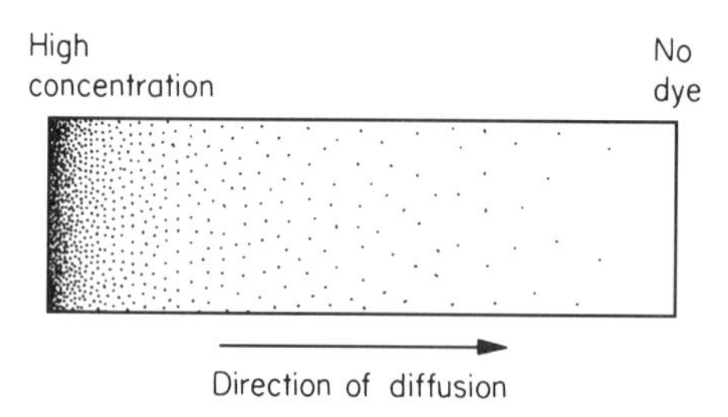

Figure 4.6 Dye diffusing in water. The dots represent the random dye molecules, and the rest of the space is filled with randomly moving water molecules (which are transparent in this sketch!).

Although Equation (4.10) was relatively straightforward to derive, it is its solution that engineers seek! For example, they may wish to know, if the inside of a nuclear reactor wall experiences a sudden increase in temperature, how quickly the temperature of the outside wall will change – or how quickly the skin of a satellite will heat up when reentering the earth's atmosphere. For a gas turbine, Equation (4.10) must be solved in order to determine how quickly heat can be conducted from the turbine blades and what the temperature distribution will be within them. For an automobile it must be solved to determine how the heat will be conducted through the engine walls. These are, in general, three-dimensional problems, whereas Equation (4.10) is a one-dimensional equation (there is only variation of T in the x direction). Its generalization to three dimensions is straightforward, and the final form is

$$\frac{\partial T}{\partial t} = \alpha\left(\frac{\partial^2 T}{\partial x^2} + \frac{\partial^2 T}{\partial y^2} + \frac{\partial^2 T}{\partial z^2}\right). \tag{4.12}$$

Now, the temperature is a function of three dimensions plus time. Finding solutions for Equation (4.12) for complicated geometries is a problem that you will deal with later, both in mathematics and heat transfer courses. We will conclude this section with a one-dimensional problem.

Consider a wall of a room that is kept at 20°C (68°F) inside. The outside temperature is 0°C (32°F). The wall is 10 cm thick and is made of solid pinewood, which has a thermal conductivity, k, of 0.15 W/(m K). Assume a steady situation such that all the heat entering the wall from the warm room is lost to the outside air. What is the transfer rate of heat to the outside (Figure 4.7)?

Because the heat flow is steady, that is, it is not changing with time, then although the temperature will vary with distance across the wall, it will not vary with time as well. Thus, Equation (4.10) becomes

$$0 = \alpha\frac{\partial^2 T}{\partial x} \quad \text{or} \quad \frac{d^2 T}{dx^2} = 0.$$

Here I have returned to the ordinary derivative notation because T is only a function of x. Integrating this equation once we find

$$\frac{dT}{dx} = \text{constant.}$$

Thus, the temperature gradient is constant across the wall (Figure 4.7), and its value is 20/0.1 or 200 K/m. To find the heat transfer rate we use Equation (4.2):

$$\dot{q}_c = -k\frac{dT}{dx} = -0.15 \times 200 = -30\,\text{W/m}^2.$$

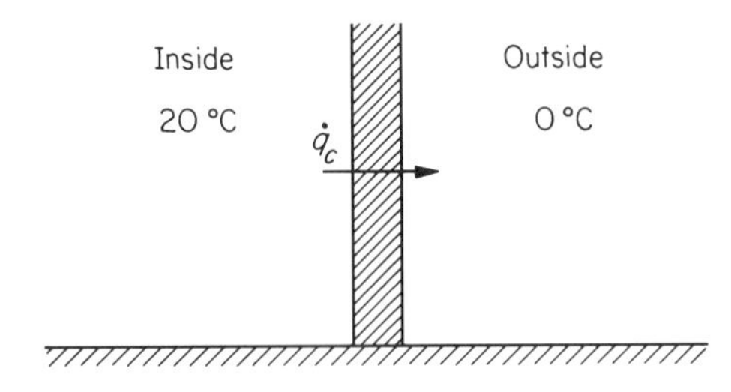

Figure 4.7 Heat conduction through a wall.

If the wall area was, say, 2 × 10 m, this would mean a heat loss of 600 W, just for one wall. Better to use insulation. Glass wool has a thermal conductivity of 0.04 W/(m K), and if the wall were made predominantly of this material its heat loss would be reduced to 160 W.

If the outside temperature suddenly got colder, the temperature gradient would increase and so would the heat loss. This is now an unsteady heat conduction problem because it would take time for the heat flow on the inside facing wall to respond to the change in conditions outside. Thus $\partial T/\partial t \neq 0$. The solution of the heat conduction equation for this problem is somewhat more complicated, but it is a mathematical, not a physical, complication.

4.2 Convection

Heat conduction can take place in any substance, be it a solid, liquid, or gas. No bulk movement of matter occurs. This is obviously the case for conduction in a solid but also applies for conduction in liquids and gases: Even if the fluid is at rest it will conduct heat, just as it diffuses matter (Figure 4.6).

Now do the following experiment. Place a glass container of water on a hot plate (Figure 4.8). Heat will be conducted through the glass to the liquid. If the amount of heat transfer is very small so that the temperature of the water does not change by much, conduction will transfer heat all the way to the top of the water. However, as we all know, soon after heating begins large-scale internal motion occurs. Warm, light blobs of water rise, and cool, heavy ones descend to replace them. The large-scale motion causes small scales also. This bulk motion is called convection. (Note that we are not considering boiling here.) Convection can also easily be observed in gases. Look at the top of a hot toaster when the sun shines through the air above it. Heat is convected by the air from the hot toaster surface upwards.

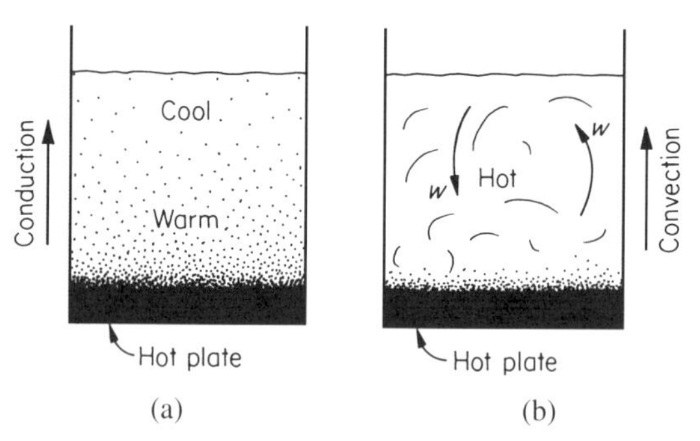

Figure 4.8 Conduction and convection. Initially (when the hot plate is just turned on) heat will be transferred by conduction, but soon convective heat transfer will take over.

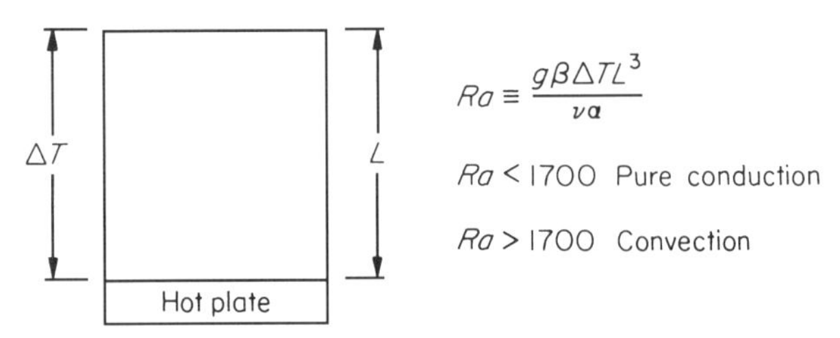

Figure 4.9 A gas in a cylinder heated from below.

Convection in liquids and gases is much more effective in transporting heat than conduction. In fact for a given temperature difference the amount of heat transferred by convection can be thousands and sometimes millions of times that transferred by conduction in the same medium over the same period. This is reminiscent of the difference between laminar and turbulent flow that we discussed in the previous chapter; the latter being much more effective in transporting momentum. They are, of course, intimately connected. If the temperature difference is great enough, the overturning motion of a fluid heated from below is in fact turbulent; in this case it is sometimes called convective turbulence. This turbulence will also cause higher stresses, and at the same time it will mix up the temperature inhomogeneities in the fluid, thus ironing out the temperature differences.

When does conduction in a fluid heated from below undergo the transition to convection? In order to address this question, we will return to our example of a perfect gas in a cylinder (Figure 4.9).

The prediction of the onset of convection from conduction is called a stability problem, just as the prediction of the onset of turbulent flow from laminar flow in a pipe is a stability problem. In Chapter 3 we showed that in a pipe, transition was determined by the flow velocity, pipe diameter, and fluid viscosity. These were grouped together to form a Reynolds number, $Re \equiv Vd/\nu$. When the Reynolds number is less than 2,000 the flow is laminar; above this value, turbulence generally occurs. A similar situation occurs for the convective instability from conduction to convection, but here you would expect different variables to be important. For example, because buoyancy is at play (hot fluid rising, cold descending), then g, the acceleration due to gravity, must be important. So too must be the coefficient of volume expansion of the fluid; as the fluid heats its density changes. For air, a perfect gas, the coefficient of volume expansion, β, ($m^3/(m^3$ K)), is $1/T$ (Problem 4.4). Large values of g and β will promote convection, as will a large temperature difference between the bottom and the top of the container, ΔT. On the other hand, if the viscosity of the fluid is very large, bulk motion will be inhibited. So too will it be if the thermal diffusivity, α, is high, because the fluid will be able to conduct large amounts of heat without undergoing transition to convection. The relevant dimensionless group for this problem is

$$Ra \equiv \frac{g\beta \Delta T L^3}{\nu\alpha}, \tag{4.13}$$

where Ra is known as the Rayleigh number. Here L is a characteristic dimension of the container. For Ra approximately 1,700, and above, convection will generally occur. (The critical value depends on the container shape, but we may regard 1,700 as a reasonable estimate for the transition.) Equation (4.13) makes sense, with large g, β, and ΔT causing a high Ra and hence convection, and large ν and α inhibiting it. The length scale occurring in the numerator also makes sense. We would not expect to see convection inside a very small container. If we consider air at 300 K, with, say, a 10 K temperature difference between the top and bottom of the container of height $L = 10$ cm, Equation (4.13) yields an Ra of 3.3×10^6. Here we have used the values $g = 9.8$ m/s^2, $\beta = 1/T = 1/300$ (K^{-1}) and $\nu = \alpha = 10^{-5}$ m^2/s. Clearly, the motion must be convective. In order to reduce the Ra below 1,700 for this situation, ΔT must be 5.2×10^{-3} K; a small temperature difference indeed. This suggests that air, when heated from below, will invariably be in convective motion, and our experience confirms this.

The transition processes from laminar to turbulent flow in pipes or from conduction to convection in heated fluids are quite complex. In all real fluids there are always small disturbances or perturbations: There are slight departures from the parabolic profile in laminar pipe flow (Figure 3.6) or small density variations at a particular height for a fluid heated from below but still in the conductive mode. Below the critical Re or Ra, these disturbances are damped out by viscous or conductive effects (or a combination of both), but above the critical value, the instabilities grow, feeding on the velocity or temperature differences. The motion becomes chaotic and then fully turbulent. An analogy is with a rock on a mountain side. If there is a wind gust or an earth tremor, the rock will be displaced and start to roll. Providing the rock is small and the friction is large, the rock will soon come to a standstill. However, if the rock is large, or if the friction is small, the rock will gather speed, dislodging other rocks and causing an avalanche, thereby changing the whole face of the mountain. Shear and convective instabilities in fluids and the onset of avalanches all involve transitions. Remarkably, similar mathematical techniques are used to describe these rather different transition phenomena. Part of the technical apparatus is called chaos theory.

Although similar mathematical techniques are used for studying transition phenomena, be they transition from conduction to convection in a container or laminar to turbulent transition in pipe flow, the actual nature of the transition processes is quite different. In pipe flow laminar to turbulent transition occurs in spots: As the Reynolds number of the flow is increased, small regions of turbulence occur in the otherwise laminar flow. These then spread throughout

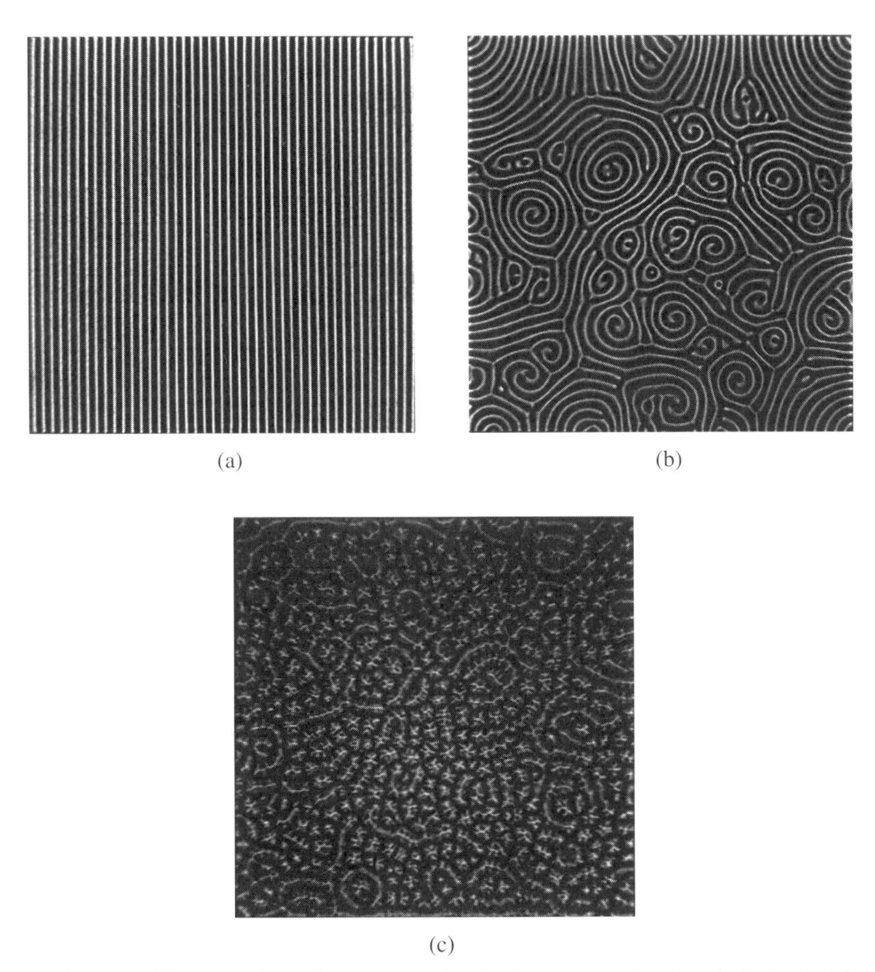

(a) (b) (c)

Figure 4.10 Top view of the onset of turbulent convection in a fluid heated from below. (a) Rolls appear at $Ra \sim 1{,}700$. (The light regions are the cold downflow.) As the Ra is further increased, (b), (by increasing ΔT) chaotic motion (called spiral defect chaos) occurs. This is followed, (c), by turbulent flow. The fluid is CO_2 (gas), and the Prandtl number is 1. (Courtesy of Professor E. Bodenschatz, Cornell University.)

the fluid as the Reynolds number is increased further. On the other hand, for a fluid heated from below, the onset of convection (at $Ra \sim 1{,}700$) is associated with orderly rolls (or under some conditions, hexagonal cells). These rolls bring warm fluid from below and cold fluid from above (Figure 4.10). As the Rayleigh number increases further, these cells become less orderly, or chaotic, and by $Ra \sim 5 \times 10^4$ the motion becomes turbulent.

Let us assume the Ra is greater than 5×10^4, that is, the convection is turbulent. (Our example of the air in a container suggests that in many practical situations the Ra is indeed much higher than 5×10^4.) In order to solve practical problems, we ask: What are the characteristic velocities (and time scales) of convective turbulence? As for the mechanical turbulence, discussed in Chapter 3, we use simple, order of magnitude analysis. Here the characteristic velocity must be a function of β, g, ΔT, and L. We have excluded ν and α because we are now assuming the Ra is high and that the damping caused by the viscosity and diffusivity is unimportant. The same idea was used when we determined the characteristic velocity of mechanical turbulence in Chapter 3. The quantities of β, g, ΔT, and L can be grouped to yield a velocity:

$$w \sim (g\beta L\Delta T)^{1/2}. \tag{4.14}$$

We have used the same symbol, w, as we did for the mechanical turbulence because both are turbulent eddy motion. You should check that the right-hand side of Equation (4.14) has dimensions of velocity and convince yourself that no other combination will do. Using the numerical values from the heated cylinder example, with $\Delta T = 10$ K, we find $w \sim 0.18$ m/s, or approximately 7 in./s. This is the characteristic turbulence velocity in the container. Watching convective motion in air, such as above a toaster or radiator, suggests this is a reasonable estimate. The time scale of this motion is calculated in the same way as for mechanical turbulence as $\tau \sim \ell/w$ (Equation (3.43)). Assuming $\ell \sim L$, that is, that the largest eddies are the same size as the container (10 cm), then $\tau \sim 0.1/0.18$ or approximately 0.5 s. This is the approximate lifetime of the eddies in the container and it is also the approximate time it takes to convect the heat so that the temperature difference between the top and bottom of the container is evened out.

We can now calculate the *relative* mixing effects of conduction and convective turbulence. Equation (4.10) shows that for conduction, the rate at which the temperature changes, for a given temperature distribution, is proportional to α, the thermal diffusivity. Now, α has units of m^2/s, which is a velocity times a length. These, for a gas, are approximately the average velocity of the molecules and the average distance between collisions respectively. This seems reasonable because for conduction the random molecular motion is responsible for the transport of the heat from high to low temperatures. Now, when there is turbulence, the large-scale random motion effectively produces a new diffusivity, which is called the *thermal eddy diffusivity*. However, we must be careful not to be too literal with our comparison. After all the thermal diffusivity, α, is a property of the fluid, just as is its density or temperature, whereas the thermal eddy diffusivity is a function of the motion; it will vary with container size, ΔT, and so on. For the gas cylinder example the thermal eddy diffusivity, which we will write as α_T, can easily be determined. Because its dimensions must be m^2/s,

$$\alpha_T \sim w\ell. \tag{4.15}$$

Notice that because w is a function of g, β, and ΔT, (Equation 4.14), so then is α_T. For our cylinder example, $w \sim 0.18$ m/s and $\ell \sim 0.1$ m, so $\alpha_T \sim 1.8 \times 10^{-2}$ m^2/s. Now the thermal diffusivity, α, for air is approximately 10^{-5} m^2/s, so $\alpha_T/\alpha \sim$ 2,000. So for this problem the heat transfer will be 2,000 times more effective when convection occurs. That is, for the same temperature difference the convection will transport the same amount of heat as conduction, but in less than one thousandth of the time. We determined that when convection occurs, the characteristic mixing time was approximately 0.5 s. If conduction were the only mechanism by which the heat was transferred throughout the fluid, the mixing time would then be 1,000 s or over 16 min. In larger systems such as the atmosphere or the sun, convection is millions and sometimes billions of times more effective than convection. For example, if the only mechanism for evenly heating a large room from a space heater placed in one corner was by conduction, it would take days! However if convection occurs the characteristic heating time is of the order of minutes (see Problems 4.5 and 4.6).

Fortunately, convection occurs in most engineering problems because we have shown that the Rayleigh number is invariably high. But consider what occurs if the fluid is not heated from below, but from above. For this case the heavy fluid is at the bottom, where it wants to be, and no convection takes place. When this type of condition occurs in the atmosphere, it is called a temperature inversion (Figure 4.11). Vertical mixing is mainly by means of conduction, and this is very inefficient, so there is virtually no vertical transport of heat or pollution. We will return to this very important problem in Chapter 5.

We have been considering only free convection, where the turbulence is caused solely by the heating of the fluid from below. In many engineering and environmental situations, there is a mean wind or fluid flow past the hot object

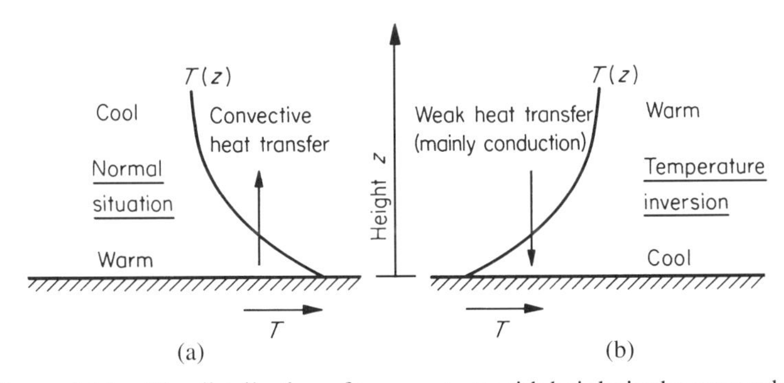

Figure 4.11 The distribution of temperature with height in the atmosphere under normal (daytime) conditions, (a), and under inversion conditions, (b). For the inversion, the cold (heavy) air is at the bottom (the opposite to convection) and hence there is no tendency for vertical air motion. Severe pollution episodes can occur under these conditions.

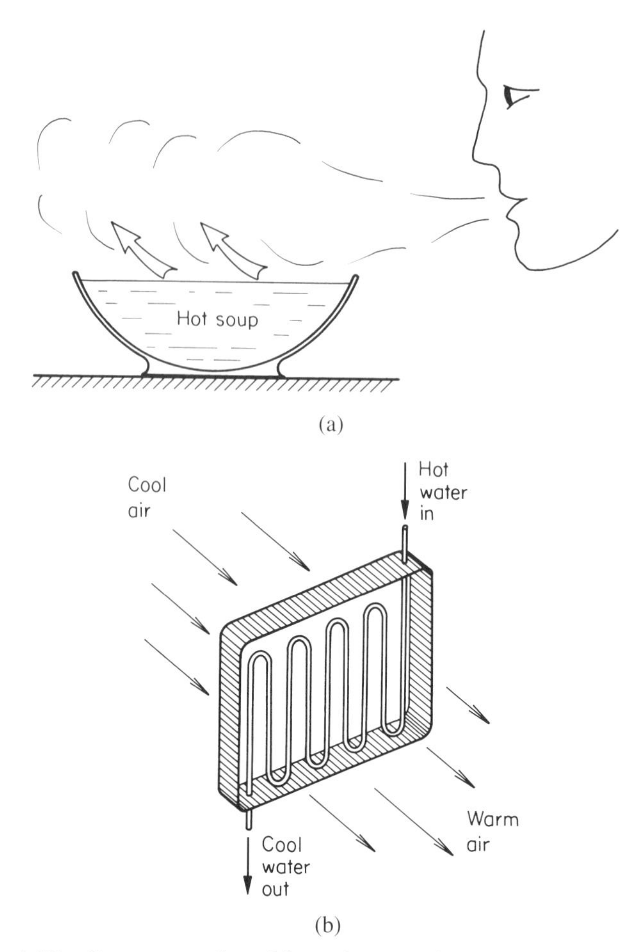

Figure 4.12 Some examples of forced convection.

(Figure 4.12). This flow itself is usually turbulent, because we have shown in Chapter 3 that most engineering and geophysical flows occur at high Reynolds numbers. This turbulence itself is an excellent mixing agent for the heat coming from the object. Thus, even if the Rayleigh number is very low, the turbulence of the incoming stream will cause efficient cooling (or heating if required). This is called **forced convection**. Here buoyancy forces can be neglected because they are small compared to the force of the motion (inertia forces). For this problem the concepts of mechanical turbulence outlined in Chapter 3 can be used. Often free and forced convection occur together (both the Re and Ra are high). This is the case in the lower atmosphere, where there is usually a mean wind over the warm ground. But there are some situations where both the Re and Ra are very low, and cooling presents a major problem. This presently is the case with some computer circuit boards. While each memory or logic chip only generates

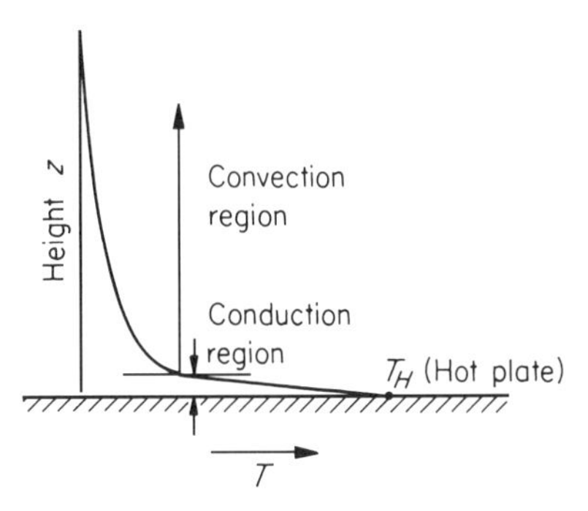

Figure 4.13 For a very thin region near the surface, the heat transfer is by conduction, even for high *Ra*, when the rest of the heat transfer is by convection.

a small amount of heat, there are often thousands of them confined to a very small space. The very small characteristic dimensions, coupled with the low air speeds that can be attained in these confined spaces, produce low heat transfer rates and this can result in the chips heating to critical levels and burning out.

At the beginning of Section 4.1 I mentioned that conduction and convection were often linked. In fact, although it is possible to have pure conduction in a fluid, when convection occurs conduction is always present. First, at the walls there is always a very thin layer of fluid that remains dominated by viscous effects. Here, because of the no-slip condition, the local velocity is small and the flow is essentially laminar. The velocity close to the wall is matched to that at the wall. Here too there is a matching of the wall temperature to the temperature of the fluid (Figure 4.13). Thus, although in the body of the fluid convection may have been very effective in mixing the hot fluid from below with the cool fluid above, producing a temperature profile that is nearly uniform (Figure 4.13), a thin region with a sharp temperature profile always remains. Here conduction effects dominate.

Second, in the convective motion itself, conduction also occurs. In Chapter 3 we saw that turbulence consists of a broad range of scales. For scales smaller than the smallest scale (the Kolmogorov scale, η, Section 3.7), the turbulent eddy motion is damped out. Because convection is a form of turbulence, it too has a broad range of scales extending down to the Kolmogorov scale. In the engine example η was around 0.1 mm. In the atmosphere it is around 1 mm. Within these small eddies, the sole mechanism for distributing the temperature differences is conduction, because there is no turbulence. Thus, it is conduction that smears out the very smallest temperature differences that the turbulence, which is occurring at the larger scales, cannot deal with. If conduction were absent, when you stirred the bath, making it turbulent and thereby mixing the cold and hot water, you would still find very small pockets of water with sharp temperature differences (Figure 4.14). Similarly, if molecular diffusion were not present, when

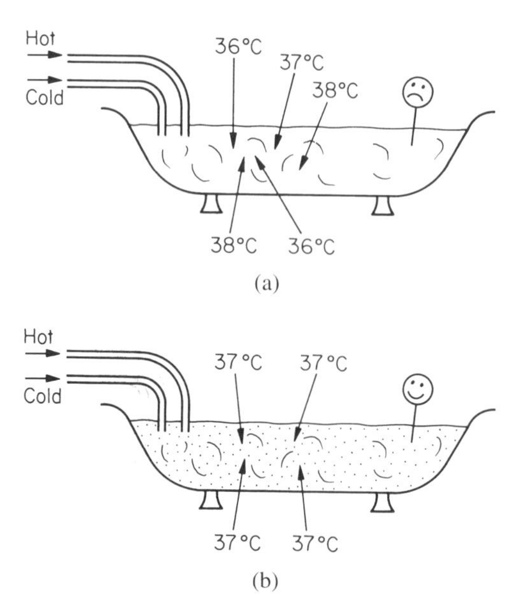

Figure 4.14 When heat transfer is by convection (be it free, forced, or mixed convection), there are still very small regions in the fluid, at scales less than the Kolmogorov scale, where temperature differences must be evened out by conduction. Fortunately, because these scales are small and because the convective turbulence has mixed up the large temperature differences, the conduction process is rapid. The bath water in (a) has been largely mixed by convection, but the fluid is imagined to have no diffusivity. In (b), the diffusion process has eliminated the small remaining temperature differences.

you stirred the milk into the coffee a fine graininess would always be present. Molecular diffusion evens this out.

We have sketched here only the basis of convective heat transfer. As you can imagine, the details of the convection will be different when it takes place over a stove top or on the surface of a heated vertical wall or in the atmosphere. Similarly, in a confined space, the large-scale convective motions will be different from those in an open container. Such problems are very important from an engineering viewpoint. Unfortunately, they are invariably too difficult to solve exactly, for similar reasons to the turbulence problem itself, and engineers have had to resort to empirical and approximate methods. They present their results in terms of the heat transfer coefficient defined by the following expression

$$\dot{q}_w = H_c(T_w - T_\infty). \tag{4.16}$$

Here $\dot{q}_w$ is the rate of heat transfer from the wall to the fluid (W/m^2), H_c is the heat transfer coefficient W/(m^2 K), and T_w and T_∞ are the temperature at the wall and very far from it respectively. In order to determine $\dot{q}_w$, H_c must be known. But H_c is a function of the nature of the fluid flow. Its value will be different for laminar or turbulent flow, or for free or forced convection, or for a mixture of both. It will also be a function of the flow geometry, having different values for a flat plate or for a cylinder, for example.

Determining H_c lies at the heart of heat transfer research, just as determining the stress at the wall lies at the heart of much of fluid dynamics research. Determination of either relies on an understanding of the fundamental structure of the fluid flow itself. We will conclude this section by discussing the relationship between the heat transfer coefficient and the fluid motion, for forced convection over a flat plate. You may think of a fast wind blowing over a warm road or someone blowing across a large bowl of thick hot soup, such as is shown in Figure 4.12. (The soup must be thick. If waves occur due to the blowing, the problem is much more difficult!)

Because for forced convection buoyancy forces are negligible, the heat transfer coefficient should be independent of such quantities as g and β. This suggests that for a particular fluid, the heat transfer should be a function of the flow conditions only. For example, if we blow slowly it should be different than if we blow fast. In particular, if we blow very slowly so that the heat transfer is laminar, we would expect quite a different dependence of H_c on flow conditions than if the flow were turbulent. (Remember that if the flow is slow (particularly if it is laminar) ΔT, the difference in temperature between the surface and the fluid, should be small. Otherwise free convection could occur. See Problem 4.9.) These considerations suggest that for forced convection, H_c should be a function of the Reynolds number only and that for laminar flow, the functional dependence should be different from that for turbulent flow. In order to relate H_c to Re we must define a nondimensional heat transfer coefficient. This is called the Nusselt number and is defined as

$$Nu \equiv \frac{H_c L}{k}, \tag{4.17}$$

where L is the length of the surface being cooled and k is the thermal conductivity of the fluid. For laminar forced convection for a particular fluid, it can be shown by exact analysis that

$$Nu \propto Re_L^{1/2}. \tag{4.18}$$

For turbulent forced convection, experiments show that

$$Nu \propto Re_L^{4/5}. \tag{4.19}$$

Here $Re_L \equiv VL/\nu$. For air the proportionality constants are approximately 0.6 and 0.03 for the laminar and turbulent flow respectively. Equations (4.18) and (4.19) are important because they enable us to determine the rate at which a hot surface is cooled if we know the Reynolds number of the flow.

In practice there may be a combination of free and forced convection, for example, a moderate breeze over a hot roof. Here g and β become important and the empirical relations for the heat transfer coefficient become more complex. Sometimes, too, there may be a combination of laminar and turbulent flow. For example, at the leading edge of the soup bowl, the boundary layer will be thin and the flow will be laminar. As the distance increases, the boundary layer thickens and the flow becomes turbulent. (See Problem 3.2.) Here too the relationship between the *Nu* and *Re* is more complex. Our understanding of forced convection is far from complete, and because of its importance it is studied widely, both in universities and in industry.

There is also much research being done on how the Nusselt number varies as the Rayleigh number is varied in free convection. As we have previously mentioned, for fluid in a container, the transition from conduction to free convection occurs at $Ra_L \sim 1{,}700$ (where here L is the depth of the container). By $Ra_L \sim 5 \times 10^4$ the motion is fully turbulent, yet recent research shows that there are other more subtle transitions that occur at much higher Rayleigh numbers. These change some characteristics of the turbulent convection. Using gaseous helium at very low temperatures, a variation of Ra from 10^7 to 10^{13} has been achieved in a small cell (of height 9 cm) by varying the viscosity and diffusivity of the fluid. Insight into the nature of the convection inside the sun and stars can be gained from these types of experiments.

4.3 Radiation

Finally, we come to the subject of radiation, the third way in which heat transfer can occur between bodies of unequal temperature.

All matter radiates energy, regardless of its composition, shape, or whether it is a liquid, gas, solid, or plasma. Radiation is due to the emission of photons resulting from the oscillation of charge within the atom and occurs for all matter above 0 K (−273.2°C).* When a body radiates it loses energy, and you may protest that after a time it will wind down to 0 K, losing all its energy as radiation. However, it continuously receives radiant energy from its surroundings so that an equilibrium temperature is achieved. Because the radiation is produced by the atom or molecule itself, it does not require a medium and can travel through a vacuum, as is approximately the case for most of the journey of the radiation from the sun to the earth. On the other hand conduction and convection require a medium.

Radiation is vital to our understanding of the atmosphere, because most of the energy the atmosphere receives is, directly or indirectly, from the sun. For

* The only exception appears to be black holes. Here the gravitational field around the black hole is so intense that radiation cannot escape. This exception to our assertion that all bodies radiate will not affect our study of the engine and the atmosphere.

the engine, an understanding of radiation is less vital because here most of the heat transfer is due to convection and conduction. However, at the spark itself, radiative effects can be important. Of course there are many other areas of mechanical engineering where radiative effects may become dominant, such as in combustion chambers, on the blades of turbines operating at high temperatures, and in the insulation of buildings. When we study the greenhouse effect we will be most interested in the effect of the atmosphere on radiation. But first we will discuss radiation through free space.

4.3.1 Radiation in a Vacuum

While radiation can be described in terms of particles called photons, it is equally appropriate to describe it in terms of waves. From this viewpoint there is no fundamental difference between X-rays (wavelength of approximately 10^{-10} m) and electric power (wavelength of approximately 10 km), although the way they are generated and the way they interact with matter is different. The distribution with wavelength of radiant energy, or the electromagnetic (e-m) spectrum, is shown in Figure 4.15. You may wish to think of it as a piano keyboard, with long wavelengths such as radar and electric power represented by the bass end, and gamma rays and X-rays by the high-pitched treble end of the keyboard. Of course, the thermal (infrared) and visible portions of the e-m spectrum have particular importance for us because radiation at these wavelengths directly affects our senses. On the other hand, we hear all the notes of a piano keyboard, and so there is no one interval that is more special than another. But consider an animal that hears only a few notes, say in the treble part of the keyboard. For it, these few wavelengths would assume a much greater importance than the rest of the keyboard, assuming it had enough reason to ponder the issue!

Because radiation is a wave, we can ascribe to each wavelength a frequency.

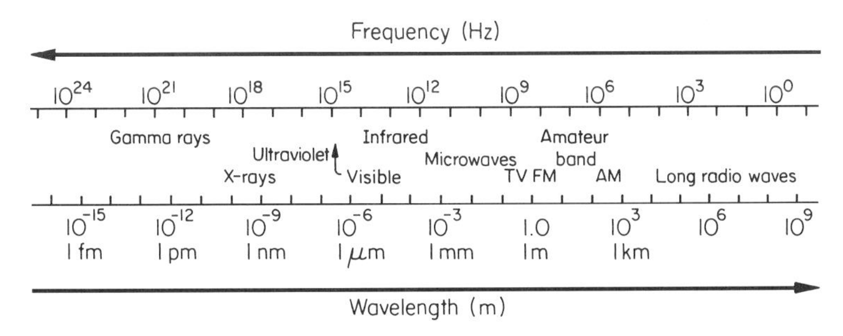

Figure 4.15 The electromagnetic spectrum. The visible region (0.4 to 0.7 μm corresponds to frequencies varying from 7.5×10^{14} to 4.3×10^{14} Hz).

The wavelength and frequency are related by the expression

$$\lambda = c/f, \tag{4.20}$$

where λ is the wavelength (m), f is the frequency (Hz), and c is the velocity of light. It is very close to 3×10^8 m/s in a vacuum and slightly less than this in gases.

The relationship between the power radiated by a body at a particular wavelength and the temperature of the body is given by Planck's radiation law:

$$E_{b\lambda} = \frac{A}{\lambda^5[e^{B/\lambda T} - 1]}, \tag{4.21}$$

where $E_{b\lambda}$ is the power radiated at a particular wavelength (W/m^2 per wavelength), T is the absolute temperature (K), and $A \equiv 2\pi h c_0^2$ and $B \equiv hc_0/k_B$. Here h is Planck's constant (6.626×10^{-34} J s), c_0 is the velocity of light in a vacuum, and k_B is Boltzmann's constant (1.3805×10^{-23} J/K). This law is the cornerstone of radiation theory and ushered in the subject of quantum mechanics at the beginning of the twentieth century. It applies to ideal or **black-body** radiators, which are perfect emitters and absorbers of radiation. (Real bodies may reflect or transmit radiation; these will be discussed below.) Figure 4.16(a) is a plot of Equation (4.21) for two black-body radiators, one at 5,770 K and the other at 255 K. (These are rather special radiating temperatures because they are the effective radiating temperatures of the sun and earth respectively.) Because A and B are constants, the form of $E_{b\lambda}$ is relatively simple; it is only a function of T and λ. Note that it is not symmetrical; it is skewed to the right. It is quite clear from Equation (4.21) or Figure 4.16(a) that a body at a particular temperature radiates a broad spectrum of frequencies or wavelengths, but when the temperature is higher, the spectrum is shifted to shorter wavelengths. The most power in the spectral distribution for a body at 5,770 K is about 0.5 μm, which is in the visible part of the spectrum, whereas for the body at 255 K, the peak is about 11 μm, which is in the thermal radiation or infrared region.

Because Equation (4.21) gives the power radiated per unit area per unit wavelength, its integral with respect to wavelength should give the total power (W/m^2) radiated by a black body. This is

$$E_b = \int_0^\infty E_{b\lambda} d\lambda. \tag{4.22}$$

The result of the integration, which is straightforward but a little messy, is

$$E_b = \sigma T^4, \tag{4.23}$$

where $\sigma = \pi^4 A/15B^4 = 5.67 \times 10^{-8}$ W/m^2 K^4. Equation (4.23) is known as the Stefan–Boltzmann law, and σ is the Stefan–Boltzmann constant. It was determined experimentally by Stefan in 1879, but its form can be predicted

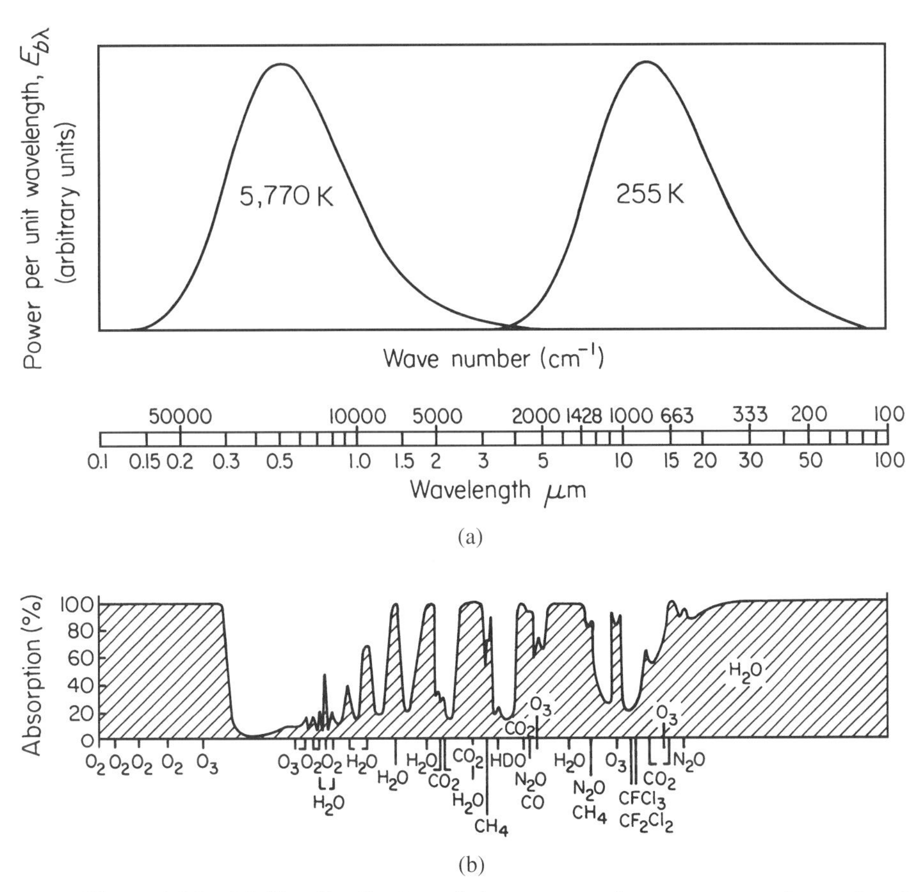

Figure 4.16 (a) The distribution of electromagnetic power per unit wavelength for the sun and the earth, each behaving as ideal black-body radiators. (b) The percentage of radiation absorbed when the sun shines through the whole atmosphere to the ground and when the earth radiates through the whole atmosphere to space (see Section 4.3.2). (After Mitchell, 1989.)

from classical thermodynamics as was done later by Boltzmann (1884). The integral, Equation (4.22), shows, however, that the Stefan–Boltzmann law is a direct consequence of the Planck radiation law, which can be obtained only using quantum mechanical reasoning. There are many other cases in physics and engineering where the integral of a function masks the true behavior of the phenomenon.

Equation (4.23) clearly shows that as T increases so does the net power radiated. Indeed, it increases very rapidly, as T^4. If you blush, raising the temperature of your cheeks (area 25 cm^2; each one) by one degree, Equation (4.23) shows that the increase in power radiated will be approximately 30 mW. This might just be detectable by your friend, even on a moonless night!

While the radiation of a black body is completely described by Equation (4.21), for real bodies this equation must be modified. Different surfaces have different **emissivities**. The emissivity, ϵ, is defined by

$$\epsilon \equiv E_a/E_b, \tag{4.24}$$

where E_a is the actual radiation. By combining Equations (4.23) and (4.24), we have

$$E_a = \epsilon\sigma T^4. \tag{4.25}$$

The emissivity is dimensionless and is 1 for a black-body radiation and less than 1 for real substances. It varies with temperature, direction, and the way the surface is treated. For example, ϵ is around 0.9 for concrete and about 0.1 for aluminum. The lower the value of ϵ, the less ability for the substance to radiate (or absorb) energy.

Nonblack surfaces do not absorb all the radiation incident onto them. They can reflect or transmit the incident radiation. This is shown in Figure 4.17, where incident radiation, G (W/m^2), arrives at the surface of a slab. The first law for the control volume (which is the slab) requires that

$$G = \rho_r G + \alpha_r G + \tau_r G \tag{4.26}$$

and

$$\rho_r + \alpha_r + \tau_r = 1, \tag{4.27}$$

where ρ_r is the reflectivity, α_r is the absorptivity, and τ_r is the transmissivity of the surface (here shown as a layer of thickness d). For a black body, which is a perfect absorber, $\alpha_r = 1$, and ρ_r and τ_r are both zero. For an opaque surface, τ_r is zero and we have $\rho_r + \alpha_r = 1$.

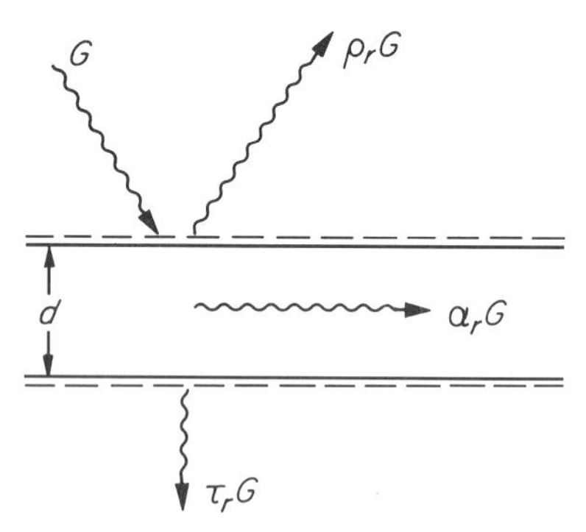

Figure 4.17 Incident radiation at a surface. Part of the radiation is reflected ($\rho_r G$), part is absorbed ($\alpha_r G$), and part is transmitted through it ($\tau_r G$).

For many substances ρ_r, α_r, and τ_r are a function of the wavelength of the radiation, and Equation (4.26) must be written in the form

$$G_\lambda = \rho_{r\lambda} G_\lambda + \alpha_{r\lambda} G_\lambda + \tau_{r\lambda} G_\lambda. \quad (4.28)$$

Here G_λ (W/(m$^2\mu$m)) is the incoming radiation at a particular wavelength, λ. For this situation,

$$\rho_{r\lambda} + \alpha_{r\lambda} + \tau_{r\lambda} = 1. \quad (4.29)$$

It is the details of the engineering rather than the basic physics that can make radiation problems very complex. For example, a hot object in a cylinder will radiate outwards, but the walls will also radiate, both back to the object as well as to the other parts of the walls. Different parts of the system may have different emissivities, and these in turn may be dependent on both the temperature and angle of the incoming radiation. All this is interactive: the incoming radiation will change the actual nature of the outgoing radiation. A simplification occurs if we consider a body of constant emissivity, and area A_1, completely enclosed by a very large surface, A_2, where $A_2 \gg A_1$. For this case we may write

$$\dot{Q} \text{ (from 1 to 2) } = \epsilon_1 \sigma A_1 (T_1^4 - T_2^4). \quad (4.30)$$

Here we are assuming the radiation is into an infinite medium of constant temperature. Consider, for example, an apple (diameter, 10 cm; temperature, 27°C; emissivity, 0.4) suspended in a large cool storage room of temperature 10°C. Equation (4.30) shows that the total radiation from the apple (assuming it is spherical) is $0.4 \times 5.67 \times 10^{-8} \times \pi \times (0.1)^2[300^4 - 283^4] = 1.2$ W. Not much, as you would expect. But if this were a steel sphere at 500°C, (with the same diameter and emissivity as the apple), the total radiation would be 250 W, about the same as a large light bulb. On the other hand, if the apple or steel ball were in a small box (slightly larger than the object), we would have to take into account the reradiated energy from the walls back to the object. We will have to take such considerations into account when we study the greenhouse effect in Chapter 5.

4.3.2 Radiation through Gases

Gases, like all matter, must radiate and absorb electromagnetic energy. When this occurs, the energy level of the molecules themselves must change, increasing when it absorbs and decreasing when it emits. The relation between the energy level, ΔE, and the wavelength of the absorbed radiation is given by

$$\Delta E = hc/\lambda. \quad (4.31)$$

Notice that the shorter the wavelength of the radiation, the larger the change in energy level. The details of the absorption depend on the molecular structure itself. For example, symmetrical molecules with two atoms (diatomic gases) such as oxygen (O_2) and nitrogen (N_2), the principal constituents of the atmosphere, can absorb or radiate energy only by changing their electronic state. This requires a large amount of energy, and hence Equation (4.31) shows that short wavelengths will be needed in order to produce the high ΔE. On the other hand, water (H_2O) and carbon dioxide (CO_2) are asymmetric molecules and can absorb or transmit energy by changing their vibrational or rotational energy. This occurs at longer wavelengths. The essential point is that each molecule in a gas can be thought of as a potential absorber or emitter of electromagnetic energy. In order to do this they have to be excited at a particular frequency or wavelength. Absorption and emission of radiation by gases occur inside combustion chambers as well as in the atmosphere. You might imagine that radiative effects combined with chemical reactions, convection, and conduction inside a combustion chamber must be very complex. In fact there are hundreds of chemical reactions going on at once, and these are affected by the radiation and turbulence levels. We will discuss this further in Chapter 6. Here we will confine our discussion to the radiation characteristics of the atmosphere because they play a central role in the greenhouse effect, to be discussed in the next chapter.

The left side of Figure 4.16(b) shows the percentage of radiation absorbed by the atmosphere due to the sun, whereas the right side of the figure shows the percentage of radiation absorbed by the atmosphere due to the radiation it receives from the earth. We have placed this figure directly below the black-body curves for the sun and the earth so you can see where the important absorption bands are relative to the emission. For example, all the sun's radiation shorter than about 0.3 μm is absorbed by O_2 and O_3 molecules. You may be aware that radiation at these wavelengths (ultraviolet, UV, radiation) is a cause of skin cancer. Although there is little fear of the atmospheric O_2 being reduced by combustion and pollution, the levels of O_3 have been reduced in recent years, allowing harmful UV radiation to reach the ground. We will discuss this further in Chapter 7. Notice (Figure 4.16(b)) that the atmosphere is almost completely transparent to solar radiation at the visible part of the spectrum (0.4 to 0.7 μm) but becomes opaque at its longer (infrared) wavelengths because of the presence of water vapor and CO_2 molecules. The evolution of our own electromagnetic radiation detectors (our eyes) in the 0.4- to 0.7-μm range is due to the transmission window in this range.

The earth too radiates to space, and here we see that there are less windows; much of the outgoing radiation is absorbed by water vapor and other complex molecules, some natural such as CO_2, CH_4, and O_3 and some products of our industrialization such as CF_2Cl_2 and $CFCl_3$ (chlorofluorocarbons, or CFCs). Here too we should include CH_4, CO, N_2O, CO_2, and O_3 because these result from industrial (as well as natural) processes. We will discuss their effects and how they are produced, in Chapters 6 and 7. It is quite evident from Figure 4.16(b) that if the atmospheric concentration of water vapor, CO_2, or any other gas that absorbs outgoing radiation from the earth increases, then these gases will absorb more energy and their temperature will increase, causing warming of the atmosphere. For this reason they are called greenhouse gases. However they too reradiate some of their energy. Part of this radiation goes to space, and part back to the ground. The situation is quite complex, but it results in a balance. When the concentrations of these gases change, then the atmosphere will have to readjust and a new balanced state will be achieved. We will investigate this in some detail in Chapter 5.

4.4 Summary

Both the design of engines and the analysis of the atmosphere require an understanding of how heat is transferred from one region to another. Here we have outlined the basic principles of the three different modes of heat transfer: conduction, convection, and radiation.

The fundamental equation of heat conduction is Fourier's Law (Equation 4.2), which states that the rate of heat transfer per unit area is proportional to the temperature gradient. In order to determine the temperature distribution as a function of position and time, an important issue in the design of turbine blades, nuclear reactors, engine blocks, and so on, we combined Fourier's law with the first law of thermodynamics and arrived at the heat conduction equation, Equation (4.10) or (4.12). Most practical engineering problems in heat conduction reduce to solving this very important equation.

Conduction occurs in liquids, gases, and solids. However, in liquids and gases different parts of the fluid are able to undergo motion relative to each other, resulting in another mode of heat transfer called convection. Convective motion can occur when a fluid is heated from below – this is called free convection – or when a flow is forced past an object that is at a different temperature – this is called forced convection. Often free and forced convection occur together. Convection usually occurs at high Rayleigh and Reynolds numbers and is much more effective in transporting heat than is conduction. However, because convective motion is nearly always turbulent, we understand it less well than we do the other two forms of heat transfer, conduction and radiation. We will show in the next chapter that convection is responsible for distributing heat, water vapor, and pollution in the lower atmosphere.

The third mode of heat transfer is radiation. Here no medium is needed. All the sun's energy reaches us by radiative heat transfer. The fundamental equation relating the energy of the radiation to its wavelength and the temperature of the emitter is Planck's radiation law, Equation (4.21). As for conduction, the basic physics is well understood; it is the details of the geometry, and the interaction of radiation with conduction and convection, that make the engineering and atmospheric problems challenging. When we deal with the greenhouse effect and global warming, we will see the very intimate connection between radiation and convection.

4.5 Problems

4.1 The temperature distribution across a furnace wall at a particular instant of time is $T = 530 - ax - bx^2$, where a is 500 K/m and b is 200 (K/m^2) and x is in meters. (The temperatures inside and outside the furnace are 530 and 298 K respectively.) The area of the wall is 10 m^2, and it is 40 cm thick. Its density, thermal conductivity, and specific heat are 1,200 kg/m^3, 25 W/(m K), and 3 kJ/(kg K) respectively. Sketch the temperature profile in the wall. Use Fourier's law of heat conduction to determine the rate of heat input to the wall ($\dot{Q}_{\text{in}}$) and its rate of heat output ($\dot{Q}_{\text{out}}$). What can you say about the rate of temperature change with time inside the wall for this temperature distribution? Can you suggest something about the nature of the temperature profile that must exist for there to be a time-varying temperature distribution?

4.2 A wall (of thermal conductivity k and area A) is bombarded with radiation, resulting in a rate of heat generation distribution $\dot{Q} = \dot{Q}_0 e^{-ax}$ inside the wall. $\dot{Q}_0$ and a are constants. Plot this distribution. Determine the temperature distribution across the wall if the hot face is at temperature T_H. (This exponential distribution of heat in the wall would be typically caused by some form of nuclear radiation that can penetrate some distance, with its effect decaying rapidly. It also describes the absorption of microwave energy inside foods that are being heated in a microwave oven.)

4.3 Consider a slab of material of width L in which heat is generated uniformly by resistance elements (Figure 4.18). The slab sits in ambient air of temperature T_0 and is deep enough so that end effects can be neglected.

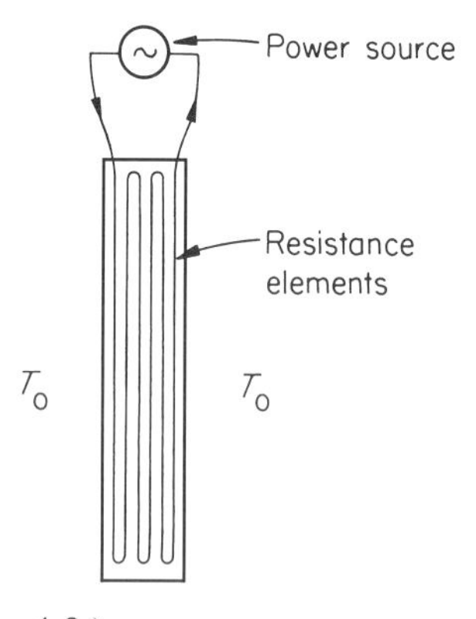

Figure 4.18 (Problem 4.3.)

The uniform strength of the heat source is P W/m^3. For this situation Equation (4.9) now becomes

$$\rho c_p \frac{\partial T}{\partial t} = k \frac{\partial^2 T}{\partial x^2} + P, \tag{4.32}$$

where ρ, c_p, and k are respectively the density, specific heat, and thermal conductivity of the slab and P is the rate of energy production. This equation states that the rate of change of temperature in the slab is equal to the sum of the rate at which the heat is diffused and the rate it is produced by the internal source. Assume now a steady-state situation ($\partial/\partial t = 0$). The problem is something like an electric blanket in a constant temperature environment. Integrate Equation (4.32) twice to determine the form of the temperature profile across the slab. Plot the profile. Does your solution make sense? (Does it have the correct symmetry properties? Are the temperatures highest and lowest where you would expect them to be?)

4.4 The coefficient of volume expansion is defined as $\beta = (1/v)(\partial v/\partial T)_p$. Show that for a perfect gas, β is $1/T$.

4.5 Consider a radiator in the middle of a room of characteristic dimension, 3 m. Although the actual surface of the radiator will be very hot, assume that the temperature a few cm above it is 10°C greater than the temperature near the ceiling (which is at 15°C).
Show that the motion in the room is convective. (Use the average temperature to determine the coefficient of thermal expansion.) What is the characteristic velocity of the motion caused by the heating? What is the characteristic time scale of this motion? Imagine now that the fluid motion in the room is completely suppressed. For this situation the heating would be due to conduction only. Its time scale would be determined by the molecular diffusivity of the air ($\alpha = 10^{-5}$ m^2/s) and the size of the room. Use dimensional analysis to estimate the characteristic diffusion time and hence the approximate heating time for this hypothetical case. How does this compare with the characteristic time scale of the convective motion? Which type of heat transfer would you prefer?

4.6 Determine the ratio of the eddy diffusivity to the molecular thermal diffusivity, α_T/α, for Problem 4.5. Your answer should be the same as the ratio of the ratio of eddy turn over time to the molecular diffusion time. (In fact these ratios will not be quite as large in practice. The vertical motion will be inhibited by thermal stratification in the room (see Chapter 5) and thus the convective velocity will be smaller (and hence the time scale larger) than your estimates of Problem 4.5. The ratio α_T/α will be about an order of magnitude smaller than your estimate. But it is still high!)

4.7 Consider an outdoor ice field of area 100×100 m^2 and surface temperature -1°C. A warm breeze of speed $V = 15$ m/s and temperature 5°C blows

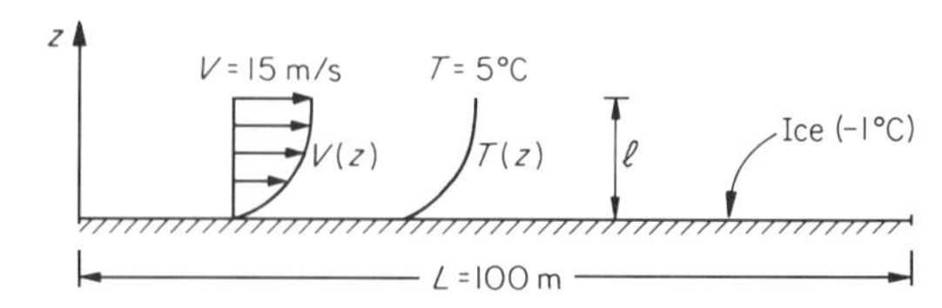

Figure 4.19 (Problem 4.7.)

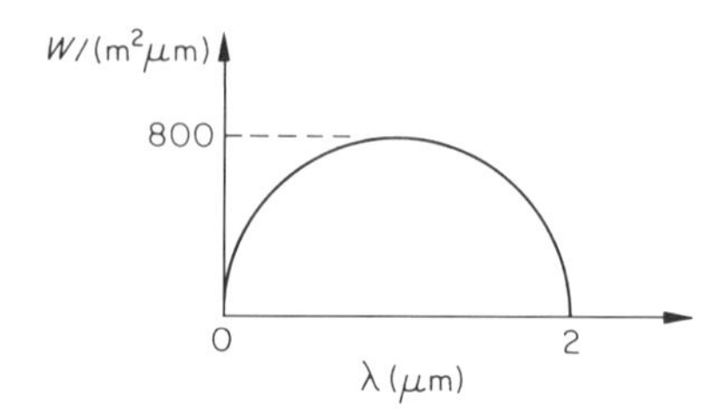

Figure 4.20 (Problem 4.10.)

across the field (Figure 4.19). Determine the Reynolds number and Nusselt number for this flow, and then determine the total cooling rate needed to keep the surface of the field at -1°C. Does your result suggest that it would be practical to build an outdoor rink of this size under these atmospheric conditions? Take, for air, $\nu = 10^{-5}$ m^2/s and $k = 0.03$ W/(m K). Assume that the flow is turbulent across the whole field, and use the Nu versus Re_L relation from the text (Section 4.2).

4.8 Show that for pure conduction, the Nusselt number is unity. (Use the Fourier law of heat conduction as well as the equations defining H_c and Nu.) What can you say about the nature of the heat transfer as Nu increases?

4.9 It is often important to determine whether a flow is predominantly free convection or forced convection. The buoyancy force per unit mass is a function g, β, and ΔT. The inertial force, or force due to the fluid motion per unit mass, is a function of V and the length scale, of the flow. Here the relevant length scale is ℓ, not L, (Figure 4.19) because it is over this length that the velocity varies, thereby creating the shear to cause the turbulence. Determine, using dimensional considerations, the acceleration per unit mass due to buoyancy and inertia respectively. The ratio of buoyancy to inertia is a dimensionless number. Compared to unity, what should this ratio be for forced convection to dominate the flow? Determine its value for the parameters of Problem 4.7. The depth of the boundary layer, ℓ, is 10 m.

4.10 A hypothetical distribution, with wavelength, of radiation emission from a surface is shown in Figure 4.20. What is the total radiation emitted?

4.11 The black body curves of Figure 4.16 have not been drawn to scale. What is the actual ratio of their peaks?

4.12 Use Planck's radiation law, with $T = 6{,}000$ K, to determine the fraction of solar radiation within the visible part of the spectrum (between 0.4 and 0.75 μm). You may use a numerical integration or any other approximate procedure. Your answer should not be surprising. Our electromagnetic sensors (our eyes) have evolved to receive radiation in this bandwidth because of its predominance.

4.13 A heating pipe of outside diameter 10 cm passes through a large room. The surface temperature of the pipe is 180°C, and its emissivity is 0.9. The temperature of the air in the room, and of the walls, is 20°C.

(a) Determine the heat transfer from the pipe (per unit length) due to radiation.

(b) Determine the heat transfer from the pipe due to convection. Assume that the heat transfer coefficient, H, is 20 W/(m^2 K).

(c) Does the relative importance of convective heat transfer to radiation heat transfer surprise you? (Have in mind a steam or hot water household radiator). Which form of heat transfer will increase more rapidly as the temperature of the radiator is increased?

Symbols

A	area	m^2
c	velocity of light	m/s
c_0	velocity of light in a vaccum	m/s
c_p	specific heat at a constant pressure	J/(kg K)
c_v	specific heat at a constant volume	J/(kg K)
E_b	total black-body radiation power	W/m^2
$E_{b\lambda}$	black-body power radiated per unit wavelength	W/m^3
F	concentration	kg/kg
f	frequency	Hz
G	incident radiation on a surface	W/m^2
G_λ	incident radiation per unit wavelength	W/m^3
g	acceleration due to gravity	m/s^2
H	enthalpy	J
H_c	heat transfer coefficient	W/(m^2 K)
h	enthalpy per unit mass (Section 4.1)	J/kg
h	Planck constant (Section 4.3)	J s
k	thermal conductivity	W/(m K)
k_B	Boltzmann constant	J/K
L	characteristic dimension	m
ℓ	turbulence length scale	m

m	mass	kg
Nu	Nusselt number	
Pr	Prandtl number	
p	pressure	Pa
$\dot{Q}$	rate of heat transfer	W
$\dot{q}_c$	rate of conductive heat transfer per unit area	W/m^2
$\dot{q}_w$	rate of heat transfer from a surface	W/m^2
Ra	Rayleigh number	
Re, Re_L	Reynolds number	
T	temperature	K
t	time	s
U	internal energy	J
u	internal energy per unit mass	J/kg
w	turbulence velocity scale	m/s
x	distance	m
z	height	m
α	thermal diffusivity	m^2/s
α_r	radiation absorptivity	
α_T	thermal eddy diffusivity	m^2/s
$\forall$	volume	m^3
β	coefficient of volume expansion	K^{-1}
λ	wavelength	m
ν	kinematic viscosity	m^2/s
ρ	density	kg/m^3
ρ_r	radiation reflectivity	
σ	Stefan–Boltzmann constant	$W/(m^2\ K^4)$
τ	turbulence time scale	s
τ_r	radiation transmissivity	
ϵ	emissivity	

5

The Atmosphere

The unifying theme of our story is the relationship between the engine and the atmosphere. Engines produce exhaust, and that exhaust inevitably ends up in the atmosphere. Hence the atmosphere is inextricably linked to the automobile and other machines that produce work. Yet there is another connection between the engine and the atmosphere: they are both complicated fluid mechanical systems at high Reynolds numbers, driven by temperature differences. The piston engine is driven by hot or cold plates (or their equivalents; see Chapter 2), and the atmospheric motion is driven by the heat sources and sinks at the equator and poles respectively. Because the motion in both systems is turbulent, predicting it and the way it mixes its various constituents (fuel in the engine; pollutants, water vapor, and heat in the atmosphere) is difficult. Of course, in order to harness the engine we control the input of fuel and air, varying it as the load varies and thereby producing a deterministic output. On the other hand, if the atmosphere can be likened to a vast engine, its output – the motion of the air masses – appears to be partly deterministic and partly random (Figure 5.1). Harnessing its power is not so easy.

The atmosphere is not only complex in its motion but also in its composition. Although oxygen and nitrogen make up over 99 percent of its mass, the very small amounts of trace gases (and solids) play a role far incommensurate with their mass fraction. Of these, water vapor is the most important, regulating the radiation balance, playing a dominant role in the atmospheric energetics by means of phase transformations, and, of course, providing the rain and snow for our very existence. Carbon dioxide, methane, ozone, and other trace gases also regulate the radiation balance and are vital for biological cycles. Finally, there are hundreds of other constituents, both natural and anthropogenic, whose concentrations are measured in fractions of parts per billion. Some of these, like carbon monoxide or bromine, play a catalytic role in chemical reactions that produce smog in the lower atmosphere and ozone destruction high in the

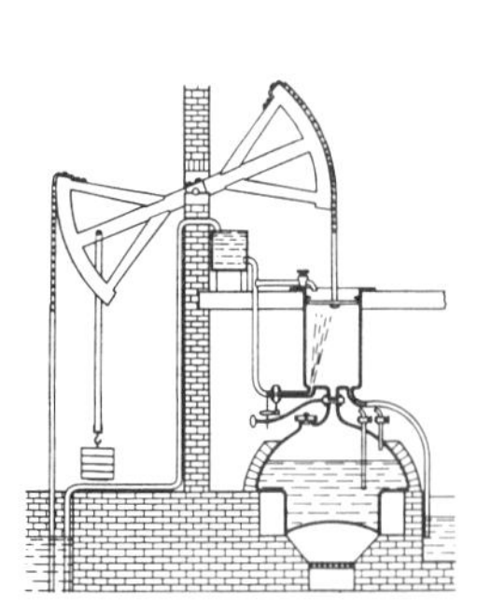

Figure 5.1 The engine and the atmosphere are both driven by temperature differences, but one produces a deterministic output, the other a combination of chaos and periodicity. (New England storm, courtesy of Mr. James Redman, Hull, Massachusetts.)

stratosphere. The atmosphere is not only a massive fluid mechanical system, it is also a chemical reactor whose richness chemists are only now beginning to appreciate.

There are many ways and levels of studying the atmosphere. The forecaster is interested in predicting tomorrow's weather; the climatologist, the next century's. The physicist wants to understand why prediction is fundamentally so difficult and whether it is fruitful to attempt to try to do better. The numerical analyst is interested in programming the equations of motion of the atmosphere in order to model it, and the planetary scientist wants to know how it got that way and why the atmospheres of other planets are different. Then there are the chemists and biochemists who want to understand the chemical reactions and their role in maintaining, or destroying, life. In all these areas engineers play a leading part. For example, atmospheric dynamics is really a problem of thermodynamics, fluid mechanics, and heat transfer. Indeed, if you visit major laboratories or university departments of meteorology and atmospheric sciences, you will find that a significant number of the researchers have come from an engineering background. An engineering degree is not useful only for designing automobiles and jet engines.

In this chapter I will focus mainly on those aspects of the atmosphere that are relevant to our theme of the engine and its relation to the atmosphere. After giving an outline of the overall structure, scales of motion, and the role of water, I will describe the greenhouse effect. It is the gases from our engines that are changing the concentrations of the greenhouse gases and thus possibly the thermal equilibrium of the planet. I will then focus on local rather than global aspects by examining the structure and dynamics within a rather small but special

Table 5.1

Atmospheric mass = 5.2×10^{18} kg
Mean earth radius = 6,370 km
Mean solar distance = 150×10^{6} km

Constituents of Clean Dry Air	
Constituent	*Volume, percent*
Nitrogen (N_2)	78.08
Oxygen (O_2)	20.95
Argon (Ar)	0.93
Carbon dioxide (CO_2)	0.035
Neon (Ne)	1.8×10^{-3}
Helium (He)	5×10^{-4}
Krypton (Kr)	1×10^{-4}
Xenon (Xe)	1×10^{-5}
Methane (CH_4)	2×10^{-4}
Nitrous oxide (N_2O)	3×10^{-5}
Carbon monoxide (CO)	1×10^{-5}

Note: 99.99 percent by volume of dry air is made up of nitrogen, oxygen, argon, and carbon dioxide

Physical Properties of Dry Air	
Density ρ at 288 K and 1,013 mb (1.013×10^5 Pa) pressure	1.22 kg/m^3
Specific heat at constant pressure c_p	1,005 J/(kg K)
Specific heat at constant volume c_v	718 J/(kg K)
Ratio of specific heats γ	1.4
(Mean) molecular weight M	28.9
Gas constant $R = R_u/M$	287 J/(kg K)

region of the atmosphere called the atmospheric boundary layer. For reference, some characteristics of the atmosphere are provided in Table 5.1.

5.1 Overall Characteristics

5.1.1 The Energy and Power of Its Motion

In Chapter 1 (Section 1.4.2) we calculated the mass of the atmosphere to be 5.2×10^{18} kg. In Problem 1.6 I asked you to determine the total kinetic energy of its motion. To do this an average speed must be assumed. I would have expected guesses to be in the range from about 1 mph (approximately 0.5 m/s) to 50 mph (22 m/s). From numerous measurements, and also by some indirect procedures, the correct value is approximately 10 m/s. In the jet stream or in a hurricane it is much faster, and on a summer evening after a hot day it is close to zero; but an average value of about 20 mph is not inconsistent with our experience. Hence the total kinetic energy of the atmosphere, $(1/2)mV^2$, is approximately 2.6×10^{20} J. In fact more detailed assessments suggest it is slightly higher, of the order of 10^{21} J. This is equivalent to the energy of over two hundred thousand, 1 megaton nuclear bombs. (One megaton of TNT is equivalent to 4.2×10^{15} J. The bomb dropped on Hiroshima in 1945 that killed 100,000 people was one hundredth of this.) A typical cyclone has an energy of around 10^{18} J, approximately one thousandth of the total kinetic energy of the atmosphere.

The atmospheric processes are driven by the energy received from the sun. The surface temperature of the sun is approximately 5,770 K, and from the Stefan–Boltzmann law ($E_b = \sigma T^4$, Equation (4.23)) you can determine that the power it generates is over 60 million W/m^2. This radiates outward, and by the time it reaches the earth (approximately 150×10^6 km), the mean power is 1,370 W/m^2. This would be the power received per square meter on a plane at the mean earth distance from the sun. If we average this power over a sphere, the value is $1{,}370/4 = 343$ W/m^2, because the ratio of the area of a sphere to that of a disk of the same diameter is 4:1 (Figure 5.2). Around one-third of this radiation is immediately reflected back into space (by clouds, snow, sea, and so forth) without playing any role in the atmospheric processes. This leaves about 240 W/m^2, or approximately 10^{17} W, for the whole of the earth–atmosphere system.

By measuring the power of atmospheric processes, an estimate of the *rate* at which the incident energy of the sun is converted into atmospheric motion can be made, and this has been found to be approximately 2×10^{15} J/s or 2×10^{15} W. This is only about 2 percent of the available power of 10^{17} W from the sun, suggesting the atmosphere is not very efficient if likened to a heat engine (Figure 5.3). What happens to the rest of the power will be explained later. Because power is energy/time, we can determine the characteristic time scale of the atmosphere as its total energy (10^{21} J) divided by its power. This is $10^{21}/(2 \times 10^{15})$ or approximately 6 days. We can think of the characteristic time scale as a kind of cycle time, or lifetime. It is the approximate time it

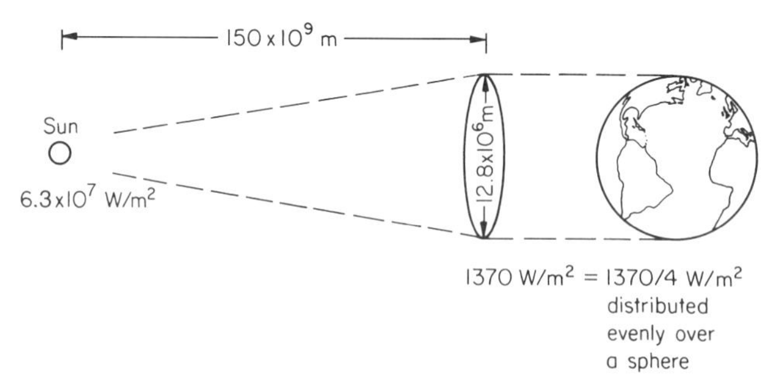

Figure 5.2 The radiation emitted at the sun, and its power per square meter intercepted at the earth.

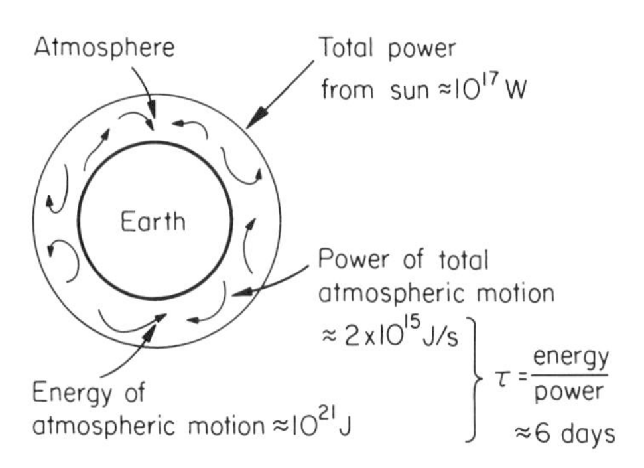

Figure 5.3 The power of the weather is only about 2 percent of the power received from the sun. Because we know both the energy and power of the atmosphere, we can determine its time scale.

takes the atmospheric motion to evolve: the time duration from the birth through the development to the demise of the very large-scale weather systems such as cyclones. We introduced the notion of characteristic time scales in Chapter 3 when we discussed turbulent eddies. Our calculated value, of the order of days, makes sense. When we look at weather maps, we see that the evolution time of the weather is on the scale of many days. It is certainly not minutes or months. Notice that if all of the sun's power were converted into atmospheric motion, the characteristic time scale would be $10^{21}/10^{17}$ or less than 3 hours. Weather forecasting would be impossible under such circumstances because entirely new weather patterns would be generated every few hours!

We could have done the above calculation in reverse. By recognizing that the time scale of the large-scale motion of the atmosphere is of the order of days, we could have determined its power, and thus the percentage of solar power converted into motion. The point is that given one or two crude estimates of, say, wind speed and characteristic time scale, many other things can be gleaned. As Mark Twain said: "There is something fascinating about science. One gets such wholesale returns of conjecture out of such a trifling investment of fact." Note also the broad-brush approach; we are not interested in fine details but orders of magnitude. Estimating the time scale as, say, 5.787 days would be quite meaningless since the atmosphere itself does not know its time scale to anywhere near this precision!

5.1.2 The Role of Water

On the average, the atmosphere contains approximately 10^{16} kg of water, mainly in the vapor state. This is equivalent to a 20-mm layer of precipitated water over the whole globe. By comparison the oceans contain over 10^{21} kg of water, five orders of magnitude greater than that of the atmosphere. The average yearly

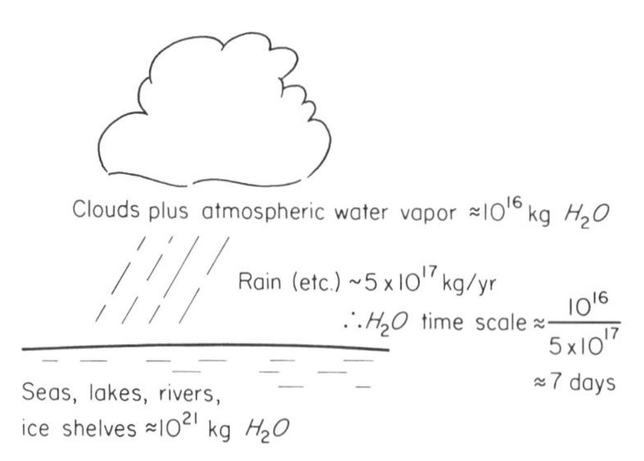

Figure 5.4 The water cycle for the earth.

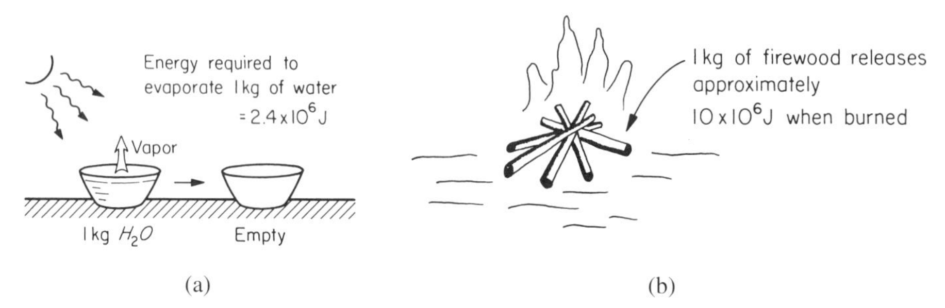

(a) (b)

Figure 5.5 The energy released when we burn a kilogram of fuel is only a few times greater than the energy required to evaporate the same mass of water. One-third of the solar energy is used for evaporation.

precipitation is approximately 5×10^{17} kg. This implies that the water vapor in the atmosphere is replaced on average $5 \times 10^{17}/10^{16} = 50$ times per year or once every 7 days (Figure 5.4). Notice here, as for the atmospheric motion itself, the characteristic time is days.

In order to evaporate water, energy must be supplied to break the intermolecular forces within the liquid (or solid if it is ice). To vaporize water, approximately 2.4×10^6 J/kg are required at normal temperatures and pressures. This is called the latent heat of vaporization or enthalpy of vaporization. It is a considerable amount of energy: the energy per kilogram released when we burn a hydrocarbon fuel is less than an order of magnitude greater than this (Figure 5.5). Because 5×10^{17} kg of water is evaporated per year, the power expended is $(2.4 \times 10^6) \times (5 \times 10^{17})$ J/year or approximately 4×10^{16} W. In the previous section we saw that the total power available from the sun is 10^{17} W. Thus, more than one-third of the sun's power is used for the evaporative process. This should be compared with the few percent used to drive the atmospheric motion. When we think of atmospheric energetics, we must be immediately concerned with the thermodynamics of the H_2O cycle.

5.1.3 Scales of Motion

By estimating the energy and power of the large-scale atmospheric motions, we determined that their characteristic time scale is a few days. We observed that this is the typical evolution time for weather maps. You will notice that when newspapers print today's weather map they compare it with yesterday's, and with predictions of tomorrow's: they do not compare it with that of the previous hour, or previous month, because for the former there would be little change, and for the latter there would be no relationship. But think of yourself as a butterfly. For your small size and weight the "weather" would be changing every few minutes. Gusts of air due to turbulence would buffet you around; a cloud would appear, changing the intensity of radiation and hence your body temperature; and a thermal plume could transport you aloft, far from your abode and possibly into the fangs of a predator, ending your brief life. For the butterfly, weather forecasts every few minutes would be desirable; daily ones would be equivalent to ten-yearly ones for humans. The time scale would be wrong.

We can study the different time scales of atmospheric processes by examining their spectrum. Such a spectrum, of the horizontal wind speed, is shown in Figure 5.6. The horizontal axis is both in Hz (cycles per second) and its inverse. The vertical axis is in units of kinetic energy per unit mass (the square of the velocity). The height of the graph indicates the energy per unit mass of the wind velocity of a particular frequency or time scale. In a similar way, the spectrum of electromagnetic radiation (Figure 4.16(a)) showed the power per unit area at a particular wavelength or frequency. The spectral representation of various quantities is used very widely in engineering and science.

The spectrum of Figure 5.6 was estimated in the following way. A meteorologist placed an anemometer for measuring the wind speed in an open field. After many days of data collection, a trace similar to that of Figure 5.7 was obtained. Here we have shown only a small portion of the trace that would go

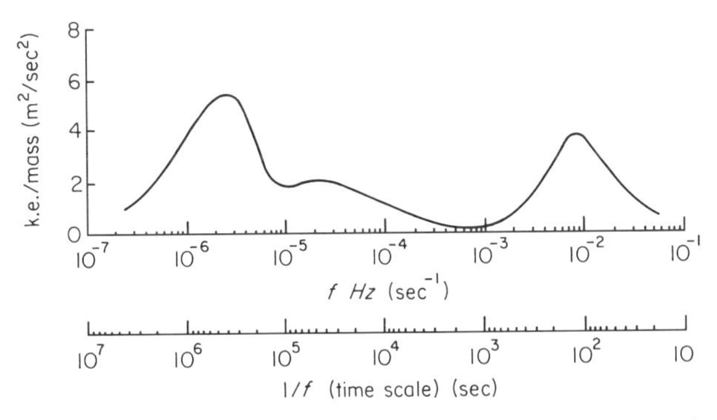

Figure 5.6 The spectrum of the kinetic energy per unit mass of the atmosphere measured at mid-latitude. (After Monin, 1972.)

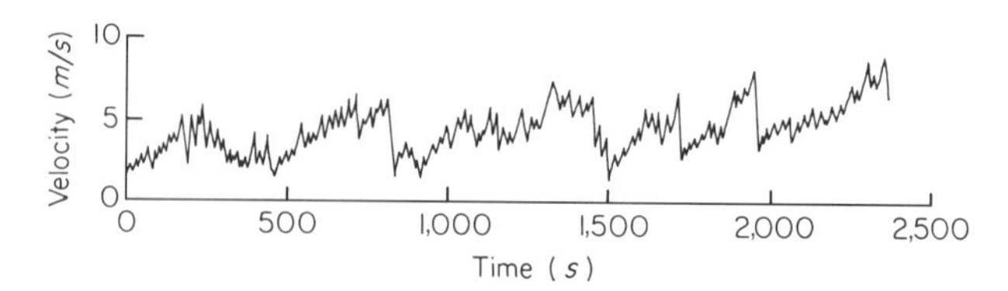

Figure 5.7 A time series from which the spectrum of Figure 5.6 was determined.

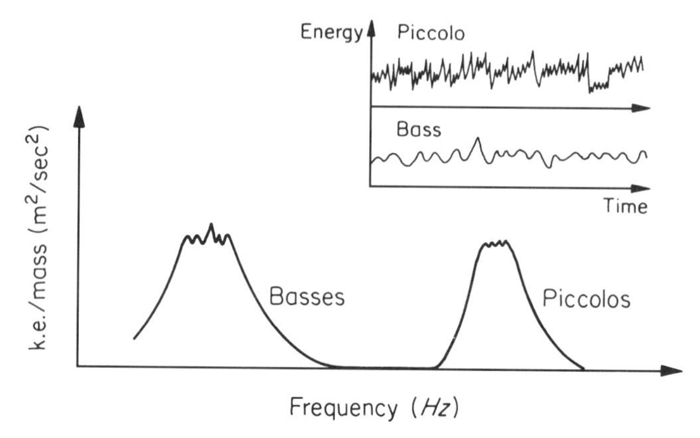

Figure 5.8 The spectrum of a piccolo and bass playing together. Compare this with the spectrum of Figure 5.6.

on for months. It consists of seemingly random ups and downs, the fast changes being due to the turbulent eddies that annoy the butterfly and blow off our hats, and the slow ones being due to the enormous "eddies" called cyclones, or what we usually refer to as weather. The graph of Figure 5.7 was then converted into the spectrum of Figure 5.6 by passing its electronically digitized signal through a vast number of filters that admit energy only at certain frequencies. The resulting atmospheric spectrum is analogous to the spectrum of musical sounds, which are the acoustic energy at certain frequencies. For example, if piccolos and basses are playing together there will be a lot of acoustic energy at the high and low frequencies but not much in the middle (Figure 5.8).

Figure 5.6 shows two broad peaks: a low-frequency peak (left-hand side) centered at about 2.5×10^{-6} Hz (4×10^5 s) and a high-frequency peak at about 8×10^{-3} Hz (125 s). The low-frequency peak has a broad extension, or lobe, on its right. Here there is energy extending to frequencies greater than 10^{-4} Hz. The low-frequency peak has the same time scale (4×10^5 s, or nearly 5 days) as that determined for the weather in Section 5.1.1. The measured spectrum has confirmed our rather crude estimate of the weather, or synoptic, time scale that was based on the overall power and energy input to the atmosphere (Figure 5.3). The high-frequency peak at about 100 s, or a few minutes, is due to the turbulent eddies. They evolve over much shorter time scales because they are small: a few hundred meters compared to the thousands of kilometers of the synoptic or large scales. Notice there is considerable energy at the time scale of 1 day (8.6×10^4 s), reflecting diurnal (day to night) variations. Notice also, there is very little energy in the region around 10^3 s (10^{-3} Hz). For these time scales the wind speed variations are not very significant.

Once characteristic energies and time scales have been estimated, the atmospheric scientist tries to determine their cause in terms of forcing mechanisms. For instance the turbulence, which is responsible for the high-frequency peak in Figure 5.6, is partly driven by the mean wind (which results from larger-scale motion) and partly by the convective motion caused by the heating of the ground. The low-frequency peak, due to the large-scale motions that we call cyclones and fronts, is influenced by the temperature difference between the equator and the poles and by the earth's rotation. Here another time scale or frequency comes into play. It is called the Coriolis parameter, $f \equiv 2\Omega \sin\phi$, where Ω is the angular velocity of the earth (7.29×10^{-5} radians s^{-1}) and ϕ is the latitude. At 40° latitude, $f = 9.4 \times 10^{-5}$ radians s^{-1} or 1.5×10^{-5} s^{-1}. We see then that the low-frequency lobe (Figure 5.6) is approximately coincident, or resonant, with the earth's rotation. On the other hand, the rotation of the earth has little to do with the high-frequency turbulence peak of Figure 5.6 because it is not resonant with it. We will discuss more about the way the smaller-scale turbulence is forced in Section 5.3.

There are even larger scales of motion than those of the synoptic or weather scales. There are global oscillations with periods ranging from weeks to months. These are due to variations of the Coriolis parameter with latitude and are associated with scales of tens of thousands of kilometers. At greater time scales than this are seasonal oscillations determined by the earth's revolution about the sun. Here it is more fruitful to examine temperature and pressure spectra rather than velocity spectra. For example, the spectrum of temperature will show broad peaks at around 24 hr and 365 days.

But there are cycles even longer than a year. There are the interannual oscillations with periods of years. For instance, there appears to be an 11-year cycle associated with the sun spots and a 26-month oscillation associated with variations in the stratosphere.

As we go to even longer time scales, our vocabulary changes. Although for periods of a few days we talk of weather forecasting; for periods of weeks and months we refer to long-term weather prediction. And as the time periods change from years to decades and millenia, the subject is called climatology. Here the issues are whether there is a temperature cycle over decades or centuries, whether the ice ages are cylical over millions of years, and if so what their forcing mechanism is. For these very long time periods our data become sketchy. There has not been enough time to complete more than a few cycles, or we do not have the long-term data. Usually it is a combination of both. Imagine, for

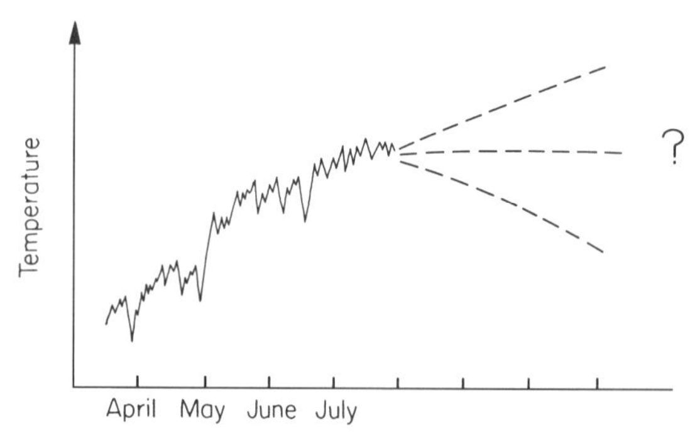

Figure 5.9 A time series of temperature measured over a few months. If you did not know that there were seasons, what would your prediction of the series be?

instance, trying to say something about the seasonal cycle if you were a strange being somehow spirited to the earth from another world and that you had only experienced the spring and summer. Your time series of temperature would show fluctuations with a gradual upward trend (Figure 5.9). If you determined its spectrum, it would have a peak for the turbulence and for the synoptic scales, but you would not be able to resolve whether there was a long-term periodicity or whether it would just keep going up and up. This is akin to the situation that we are facing with the so-called global warming issue, which will be discussed in Section 5.2.

We see, then, that in terms of its energetics and thermodynamics, the atmosphere consists of a very broad spectrum of time scales from seconds to hundreds of millions of years. The spectrum is continuous, so there must be interaction between the various scales: the motion of warm or cold fronts affects the small-scale turbulence, the seasons change the nature of the fronts, and so on. This is why the atmosphere is so difficult to predict. Think of a pendulum that, instead of oscillating with a single period, had periods ranging from seconds to years and that these oscillations were all happening together, so that its motion looked like it was being controlled by an imp. Unraveling the nature of the atmosphere is a little like working out the mind of an imp by watching the result of her actions. After a time we see some rhythms, but there will always be mysterious and unexpected happenings.

5.1.4 Its Variation with Height

So far I have been concerned only with the bulk characteristics of the atmosphere without regard to its spatial variation. There is, as we all know, pronounced variation of temperature, rainfall, pressure, and other parameters with latitude as well as with precise longitudinal location. Moscow, for example, experiences winters of great severity with temperatures often well below 0°F (–18°C), while

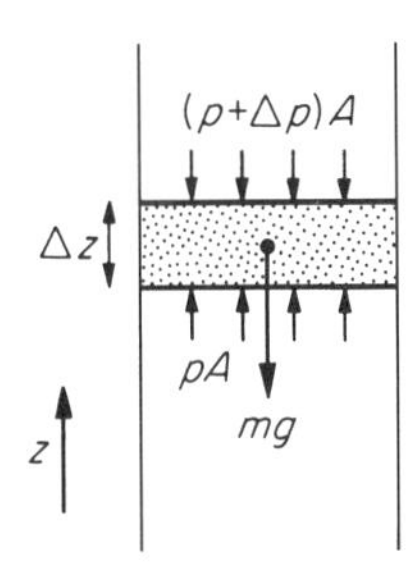

Figure 5.10 A sketch of a vertical column of air. The small pressure change, Δp, occurs over a small height internal, Δz.

Glasgow, at the same latitude, rarely has snow, and on neighboring islands such as the Isle of Aaron, lush vegetation can be found even in winter, due to the warming effects of the Gulf Stream. Weather and climate are controlled by local topography and ocean dynamics, as well as the large-scale motions that can be accounted for purely by the combination of the earth's rotation, its inclination, and the solar radiation balance. We will not be particularly concerned with the issue of topographical variation, except insofar that it is necessary to explain some aspects of engine–atmosphere interaction such as photochemical smog. However, in all pollution problems, be they local or global, it is necessary to understand the variation of the atmosphere with height, the subject to which we now turn our attention.

We begin by determining the variation of atmospheric pressure with height. Figure 5.10 shows a small volume element of the air, at rest, with the various forces acting on it. Notice that since z is increasing with height above the ground, I have assumed in Figure 5.10 that p is also. The sign of our resultant force balance will reflect our sign convention. The main point is that over the small height Δz there is a small change of pressure, Δp. Apart from the pressure forces on the upper and lower surfaces, there is also the force due to the weight of the air element. This is $mg = A\Delta z\rho g$ where A is the area of the fluid element, which has a density ρ. The force balance yields

$$A\Delta z\rho g + (p + \Delta p)A - pA = 0 \qquad \text{or} \qquad \Delta p/\Delta z = -\rho g. \tag{5.1}$$

For a differential element this may be written as

$$\frac{dp}{dz} = -\rho g. \tag{5.2}$$

Equation (5.2) is known as the **hydrostatic condition**. It can only be solved for the pressure if it is known how ρ varies with height. For example, if ρ is constant, then integration of Equation (5.2) yields

$$p_1 - p_2 = \rho g(z_2 - z_1), \tag{5.3}$$

that is, the pressure decreases linearly with height for this case. Notice that in Chapter 3 we derived this expression in quite a different way; from the Bernoulli equation. There we were studing energy balances. Here we have arrived at it from a consideration of a force balance.

The density of the atmosphere in fact decreases with height. Let us begin by assuming it decreases exponentially, that is,

$$\rho = \rho_0 e^{-z/H}. \tag{5.4}$$

Here ρ_0 is the reference (ground) value of density and H is a constant (height) to be determined. Substituting Equation (5.4) into the hydrostatic equation (Equation (5.2)), and upon intergration, we find

$$p = \rho_0 g H e^{-z/H}. \tag{5.5}$$

Since ρ_0, g, and H are constants, we see that the pressure also decreases exponentially with height. The product $\rho_0 g H$ defines the ground level pressure, p_0, and so we can write

$$p = p_0 e^{-z/H}. \tag{5.6}$$

Since p_0 is approximately 10^5 Pa, $H = p_0/\rho_0 g$ is 8.5 km. This is the height of a constant density atmosphere producing the same pressure as the exponential atmosphere at sea level (see Problem 1.7). It is known as the **scale height** of the atmosphere.

Equations (5.4) and (5.6) show that at the scale height H (about the height of Mount Everest), the pressure and density are $1/e$ or 0.37 their surface values. By 100 km, they have dropped to nearly one millionth of their values at the surface. Although the atmosphere extends to hundreds of kilometers, approximately 80 percent of its mass is in the lowest 10 km. This region is appropriately called the troposphere, or overturning sphere, because it is here that most of the clouds and complex weather phenomena occur, with associated updrafts and other overturning phenomena.

Figure 5.11 shows a plot of the variation of ρ and p with height for a standard atmosphere. It is the average of many measurements done at various locations and at various times of the year. You can show from these graphs that the exponential models (Equations (5.4) and (5.6)) are only crude approximations to the real atmosphere. Indeed Equations (5.4) and (5.6) imply that for a perfect gas atmosphere there would be no variation of temperature with height, that is, the atmosphere would be isothermal. In fact there are marked temperature variations with height, and these cause departures from the exponential distributions of ρ and p. The observed temperature distribution with height is shown in Figure 5.12. It has a zigzag structure. We use this distribution to define the atmosphere's various spheres. We have just mentioned the troposphere, the region closest to the ground. Here on average the temperature decreases with height,

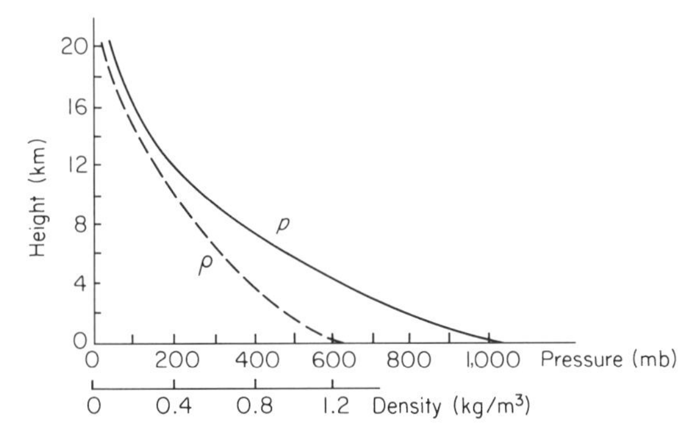

Figure 5.11 The variation of pressure and density with height in the standard (average) atmosphere. The actual values vary from day to day and season to season. Average surface values of $\rho_0 = 1.22\ \text{kg/m}^3$ and $p_0 = 1{,}013$ mb have been used (1 mb = 100 Pa).

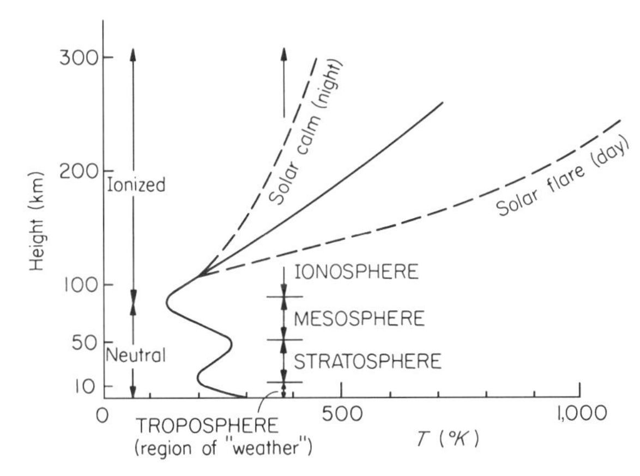

Figure 5.12 The vertical temperature profile of the earth's atmosphere. The height of the layers varies with latitude and time of year.

indicating the convective activity that we normally observe. There are, however, very important exceptions to this, as we will see in Section 5.3. The region above the troposphere is called the stratosphere. Here the temperature increases with height primarily because of the absorption of sunlight by O_3 molecules. This temperature inversion, we will show, inhibits turbulent mixing.

Up until the top of the stratosphere (the stratopause), the relative composition of the major constituents is uniform. Above the stratosphere the oxygen molecules dissociate into atomic oxygen. In this region, beginning at the mesosphere and becoming more pronounced in the ionosphere (Figure 5.12), the earth's magnetic field and the charged particles from the sun play a major role in the dynamics. Notice that here, the temperature is very high. However, because the density is very low, so is the heat capacity. This is why astronauts don't

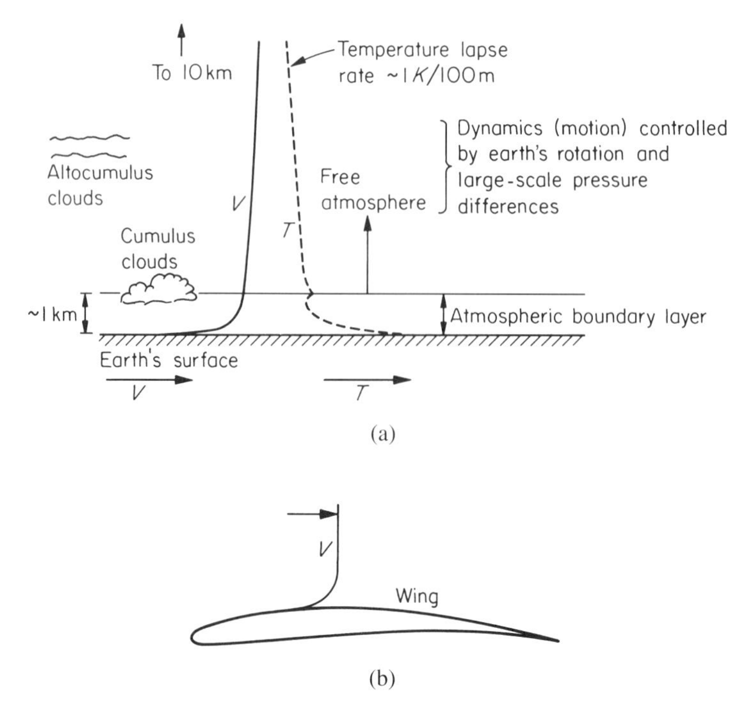

Figure 5.13 The vertical profile of the troposphere. This is an average profile. There are departures, particularly close to the ground (see Section 5.3). We have compared the atmospheric boundary layer with the boundary layer over a wing.

become baked. In terms of the effect of the engine on the atmosphere, the troposphere plays the most important role both locally and globally, although the stratosphere must be considered when discussing the ozone depletion problem, because it is here that O_3 has its highest concentration, acting as a barrier to the harmful ultraviolet radiation from the sun.

The troposphere can be further divided into different regions (Figure 5.13). From our study of fluid mechanics in Chapter 3, we saw that near a surface, a fluid is brought to rest because of the no-slip condition. The region over which the fluid velocity varies from its free stream value to zero at the surface is called the **boundary layer** (see Problem 3.2). For an aircraft wing it is only a few centimeters thick (Figure 5.13(b)). For the atmosphere it is typically 1 km, but this varies with overall weather conditions and the time of day. In the boundary layer frictional effects play a crucial role. In Figure 5.13 we have drawn the temperature decreasing with height in this region. This is for a typical day. Notice that there is a small increase in temperature at the top of the atmospheric boundary layer. This is called the capping inversion. Such inversions can also occur lower in the atmospheric boundary layer. We will discuss these in Section 5.3. The region above the atmospheric boundary layer is called the free atmosphere. Here the effects of the earth's surface (be it land or water) are less important. There is less shear, and turbulence levels are generally lower, although many of us have encountered severe clear air turbulence when in commercial airplanes. This is usually caused by local gradients in the wind velocity (wind shear). In the free atmosphere the dynamics of the air motion is controlled by a combination of the Coriolis force due to the earth's rotation and the large-scale pressure gradient, which is in turn controlled by the horizontal temperature gradients. The frictional effects due to turbulence are much less important than in the boundary layer. In fact it is in the boundary layer that most of the power of the wind, which is about 2×10^{15} W (Section 5.1.1), is dissipated into heat by the turbulence.

For the rest of this chapter I will focus on the atmospheric phenomena that are most central to our understanding of the engine and the atmosphere. I will begin by discussing the greenhouse effect because it is central to global issues, and I will then turn to local phenomena in the atmospheric boundary layer.

5.2 The Greenhouse Effect

We saw, in Section 4.3 (Figure 4.16) on radiation through gases, that the atmosphere is nearly transparent to the sun's radiation, particularly in the visible region of the spectrum, from 0.4 to 0.7 μm, where most of the sun's energy is. At shorter wavelengths O_2 and O_3 absorb the UV radiation, and at the longer (infrared) wavelengths there are absorption bands due to water vapor and oxygen. The earth, on the other hand, with an effective radiating temperature of 255 K (I will explain the reason for the below-freezing value in a moment), radiates back into space through a very absorbing atmosphere with only a few windows, the largest at around 11 μm. Water vapor, CO_2, and the other complex molecules absorb most of the outgoing radiation at other wavelengths. It is this lack of symmetry between the incoming and the outgoing absorption characteristics that is responsible for the greenhouse effect, although, as you will see, for a proper understanding other vital but more subtle aspects must be taken into account. In particular we will see that convective activity and phase changes play a most important role.

We will begin by considering the most simple model of the greenhouse effect, known as a zero-dimensional model. This does not take into account variation in any spatial direction, lumping together the earth and atmosphere as a whole. Then we will consider a one-dimensional model that examines the atmospheric and earth radiation budgets separately. In both models we will assume long-term averages, taken over a time period greater than a year, so that diurnal and seasonal variations are averaged out. The reality is of course three-dimensional with diurnal and seasonal variations, but, as we will see, at the present time not even the most advanced research can deal with this in all its detail.

The net power into the earth's climate system, $R_{net}(W/m^2)$, must be zero if we assume for the moment that it is behaving normally, that is, that no global warming (or cooling) is occurring. The solar power per unit area at the mean earth–sun distance, S_0, is 1,370 W/m^2 (Section 5.1.1). This has been established by means of satellites that measure the sun's radiation. The error is about ± 3 W/m^2 or about 0.2 percent. However, about 30 percent of the incoming solar power is immediately reflected back into space without taking part in any energetics. We say that the overall **albedo** of the atmosphere, α, is 30 percent, or 0.3. There is variation of α from year to year because it is affected by clouds, ice cover, and so forth. Thus, the net incoming radiation per unit area of the earth's surface is $(S_0/4)(1-\alpha)$. Here the factor of 4 has been used to convert from energy/area for a disk to energy/area for a sphere (Section 5.1.1).

The net incoming radiation must be balanced by the net outgoing radiation, which can be determined from the Stefan–Boltzmann law (Equation (4.23)) if we know the radiating temperature of the earth–atmosphere system, T_e. Conversely, T_e can be determined from the balance between the net incoming and outgoing radiation:

$$R_{net} = 0 = (S_0/4)(1-\alpha) - \sigma T_e^4$$

or

$$T_e = [(S_0/4)(1-\alpha)/\sigma]^{1/4} = 255\,\text{K}. \tag{5.7}$$

The freezing value of 255 K, known as the effective black-body temperature of the earth, is far below that determined by careful ground-looking satellite measurements that estimate the average temperature of the earth surface, T_s, to be 288 K (15°C or 59°F). The difference, 33 K, is due to the absorption of the outgoing radiation. If the earth had no atmosphere (but the same albedo) its surface temperature would be 255 K.

A sketch of the situation is shown in Figure 5.14. As we have shown, the net incoming radiation is $(S_0/4)(1-\alpha) = 240$ W/m^2, and so too must be the net outgoing radiation if we are to have equilibrium. This is equivalent to $T_e = 255$ K. Because the average surface temperature of the earth is 288 K, then the power from the surface is $\sigma T_s^4 = 390$ W/m^2. In the greenhouse jargon we say $390 - 240 = 150$ W/m^2 is trapped. We will now turn to the more subtle aspects of this trapping by considering the one-dimensional model.

To say that 150 W/m^2 is trapped by the atmosphere is not quite accurate. It implies that an overall warming will occur: 150 W/m^2 is a large rate of increase of energy; in no time we should all be baked. Imagine a 150-W light bulb every square meter, over the whole surface of the earth. Clearly the atmosphere

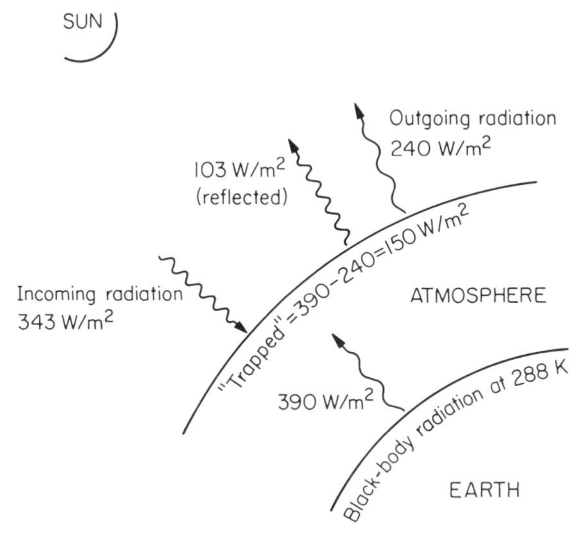

Figure 5.14 A simplified view of the greenhouse effect. The tightly bunched wiggly arrows represent incoming short-wavelength radiation. The other (less bunched) wiggly arrows represent the outgoing (infrared) radiation.

must have an equilibrium temperature too. This must be determined by its own radiating characteristics. The picture of Figure 5.14 gives only the boundary conditions: the net radiation entering and leaving the earth–atmosphere system as a whole. We must now look at the earth and atmosphere separately.

A separate balance for the atmosphere and the earth is shown in Figure 5.15. We will examine it in stages. First consider the sunlight entering. We stated above, of the 1,370/4 or 343 W/m^2 entering, about one-third, or more precisely 103 W/m^2, is immediately reflected back to space; it has no energetic interaction with the atmosphere or earth. Of the remaining 240 W/m^2, careful measurements show that most is absorbed by the earth itself (168 W/m^2). Nevertheless the atmosphere is not entirely transparent to shortwave radiation, and it absorbs 72 W/m^2. Thus, the short-wave (sunlight) radiation balance (in W/m^2) is

$$\underset{\substack{\text{total}\\\text{from}\\\text{sun}}}{343} = \underset{\text{reflected}}{103} + \underset{\substack{\text{absorbed}\\\text{by}\\\text{earth}}}{168} + \underset{\substack{\text{absorbed}\\\text{by}\\\text{atmosphere}}}{72}. \qquad (5.8)$$

This is shown on the left side of Figure 5.15.

The atmosphere too is a body, and so it must radiate energy. It is cool, so it will radiate in the infrared region. Measurements show that on average it radiates 330 W/m^2 back to earth, giving it a radiating temperature of 276K (3°C). We

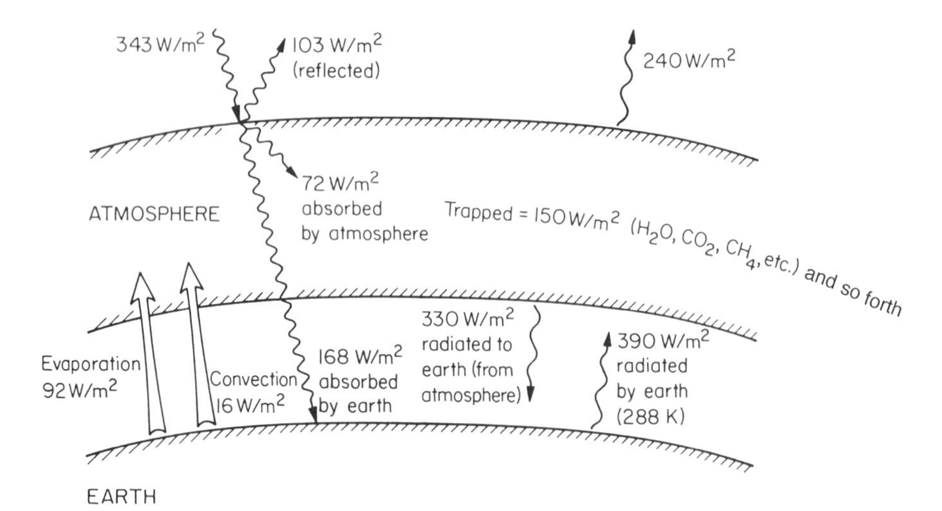

Figure 5.15 The greenhouse effect with the earth and atmospheric budgets separated. The net power into each must be zero if they are not to heat or cool. (After Ramanathan et al., 1989.)

know that 150 W/m^2 is trapped, so the radiation budget for the atmosphere is

$$\underset{\substack{\text{absorbed}\\\text{from}\\\text{sun}}}{72} + \underset{\text{trapped}}{150} - \underset{\substack{\text{radiated}\\\text{to earth}}}{330} = -108. \tag{5.9}$$

This is shown on the right side of Figure 5.15. Thus, the atmosphere is in fact losing radiative power rather than gaining it. Our calculation suggests it is in fact cooling! (Note that the 150 W/m^2 takes into account the radiative energy transmitted into space by the atmosphere because it is the difference between the incoming radiation to the atmosphere from the earth and the total outgoing radiation from the earth–atmosphere system to space.)

The radiation budget for the earth's surface is also in imbalance. It is

$$\underset{\substack{\text{absorbed}\\\text{from the}\\\text{sun}}}{168} + \underset{\substack{\text{absorbed}\\\text{from the}\\\text{atmosphere}}}{330} - \underset{\substack{\text{radiated}\\\text{to atmosphere}\\\text{and space}}}{390} = 108. \tag{5.10}$$

Here there is a net increase in radiative power. Thus, due to the radiative processes alone, the atmosphere is cooling and the earth is heating.

This radiative imbalance between the earth and the atmosphere is made up by nonradiative heat transfer processes. The most important of these is due to the energy required to cause the phase change in evaporating water. In Section 5.1.2 we saw that approximately one-third of the energy input to the earth–atmosphere system is involved in the evaporative process. More precisely, 92 W for every square meter of the earth's surface is used in evaporating water. This evaporation is a true heat transfer process, cooling the earth and warming the air aloft when the vapor condenses. The other 16 W/m^2 required to sum to the 108 W/m^2 in order to balance the budget of Equations (5.9) and (5.10) is due to convective heat transfer from the earth to the atmosphere. Thus, the earth's surface loses 108 W/m^2 and the atmosphere gains 108 W/m^2 so that their net energy change is zero.

I have been rather pedantic in explaining the principle of the greenhouse effect, not because it is so complicated but because it is different from popular conceptions. Even careful newspapers, such as the *New York Times*, are misleading when they explain the greenhouse effect by stating that the atmosphere traps radiation. In fact it loses more radiation than it gains, and its energy has to be replenished by heat addition. This is achieved by means of convection and evaporation.

We see, then, that the earth–atmosphere system is not in radiative equilibrium but that there is a delicate balance between radiation and convection. (I will lump the evaporative process with the purely convective for ease of discussion.) Altering one number in Figure 5.15 (for instance, changing the volume of water vapor or CO_2 in the atmosphere so the trapping term changes by a few percent) will cause all the other numbers to change until a new equilibrium is achieved. Thus, if the trapping term increases, the atmosphere will radiate more to the earth. The earth will then warm, and more evaporation will take place. This could cause more cloud cover and so the albedo of the earth could change, either decreasing or increasing the amount of radiation absorbed by the atmosphere, depending on the nature of the clouds. This would then cause further adjustments, and so on. It is a system of multiple feedback loops, all acting in concert.

When we consider the atmosphere–earth system in three dimensions, the situation is further complicated. Radiative power is not balanced at each latitude. The poles with their ice and snow reflect more incoming radiation than do the lower latitudes, because of their higher albedo. This increases the temperature gradient between the equator and the poles. Cloud cover varies with latitude and longitude, causing further imbalances. Clearly, a deep understanding of weather and climate, particularly the water cycle, is necessary in order to predict what will happen if there are disturbances or perturbations (such as increasing the amount of CO_2) to the earth–atmosphere system. Determining whether global change will occur is a little like a super long-term weather forecasting problem, made even more difficult by the fact that the atmospheric constituents are changing with time.

How do we go about analyzing our earth–atmosphere system in order to predict whether global warming will occur because of the change in atmospheric composition from exhaust gases and the like? We begin, as do all intelligent investigators, by looking at very simple models and then building up to the more complex. I will describe the most simple model of greenhouse warming and then outline the approach taken in contemporary research.

We are interested in the rate of change in the global temperature as the heat absorbed by the atmosphere (principally the troposphere) increases because of increases in concentration of greenhouse gases. The first law of thermodynamics for the closed atmosphere (assuming heat interactions across its boundaries and allowing for expansion) is

$$\Delta \dot{Q} = c'_p \frac{d(\Delta T)}{dt}. \tag{5.11}$$

This has been written in a rate form and per unit area; the units are W/m^2. It is similar in form to Equation (4.6). It states that a change in the rate of heat input is equal to the rate of temperature change of the atmosphere multiplied by its heat capacity. Here the units of c'_p must be $J/(m^2\,K)$, that is, the heat capacity of a $1\ m^2$ column of atmosphere, because $\dot{Q}$ is in W/m^2. (This is why I have used a superscript.) The specific heat, c_p, of air is 10^3 J/(kg K), so this translates to $c'_p = 10^7\ J/(m^2\,K)$, because the mass of $1\ m^2$ of atmosphere is 10^4 kg. (The mass of the atmosphere is 5.2×10^{18} kg, and the area of the earth is $5.1 \times 10^{14}\ m^2$.)

Now, Equation (5.11) does not properly describe the situation. It suggests that if, for instance, we add carbon dioxide to the atmosphere, thereby increasing the rate of heat absorption by a few W/m^2, then $d(\Delta T)/dt$ will be positive, that is, ΔT will increase without limit. Clearly we must take into account the fact that the atmosphere radiates long-wave radiation and that the amount of this radiation increases as the temperature of the atmosphere increases. Such a process is called negative feedback. To include it we modify Equation (5.11) as follows:

$$\Delta \dot{Q} = c'_p \frac{d(\Delta T)}{dt} + \lambda \Delta T, \tag{5.12}$$

where λ is called the feedback factor. Our interest is in determining ΔT for a given value of $\Delta \dot{Q}$, that is, for increased atmospheric absorption. Thus, Equation (5.12) must be solved for ΔT. You will soon learn general methods for the solution of differential equations such as this; here we are interested in the solution itself, which is

$$\Delta T = (\Delta \dot{Q}/\lambda)\left[1 - e^{-\lambda t/c'_p}\right]. \tag{5.13}$$

You can verify that this is correct by substituting it back into Equation (5.12). Notice that the form of Equation (5.13) is sensible. For example, for $t \to \infty$, $\Delta T \to \Delta \dot{Q}/\lambda$, that is, the increase in temperature is equal to the increase in radiation absorption divided by the feedback factor. I will ask you to further explore Equation (5.13) in Problem 5.8. Notice that if there were no feedback

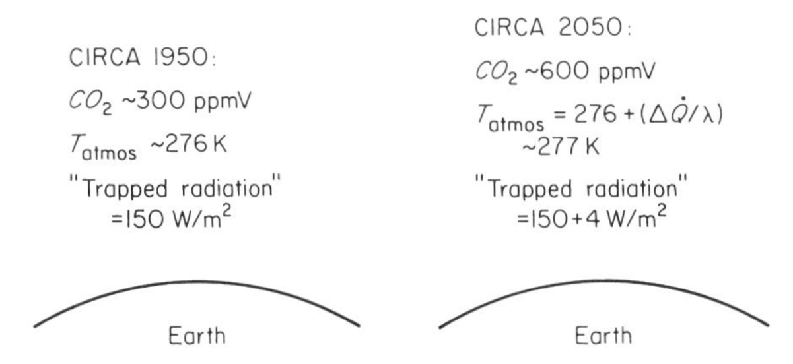

Figure 5.16 The use of Equation (5.13) to determine global warming. Here we have assumed the warming is due to an increase in CO_2 alone. In fact there are many other aspects such as the change in evaporation and cloud cover that must be taken into account.

($\lambda = 0$) then ΔT, for large t, would be infinite. This, of course, is consistent with the solution of Equation (5.11).

We can determine λ for a simple case. Consider pure black-body radiation into space (Figure 5.16). For an increase of the atmospheric temperature by 1 K (from its present value of 276 K), the amount of outgoing radiation would increase by $\sigma(T_{277}^4) - \sigma(T_{276}^4) = 4.79$ W/m^2. Here we have used the Stefan–Boltzmann law (Equation (4.23)). Thus $\lambda = 4.79$ W/(m^2 K). What about $\Delta \dot{Q}$? Detailed calculations show that if the amount of CO_2 in the atmosphere is doubled, from its value of around 300 ppmv 100 years ago (see Figure 1.6 for recent trends), then it will absorb (or trap) 4 W/m^2 more power due to absorption. Thus $\Delta T = \Delta \dot{Q}/\lambda \sim 1$K.

However, this estimate of a 1K rise in the atmosphere for a doubling of the CO_2 will be modified by further feedback trends due to changed cloud cover, increased evaporation, and so on. Each of those mechanisms will have its own λ, and its value may change as T changes. It may also change when other variables it depends on change. Different variables will play different roles with latitude and longitude. Their value will depend on whether there is land or water or whether the terrain is flat or mountainous. The list of complications appears endless and cannot be dealt with in a neat analytical way.

In fact atmospheric researchers take an approach to this problem similar to the approach that fluid dynamics engineers take in engine design; they do numerical simulations in a computer. In Section 3.7 we showed how the gaseous mixture inside an engine was divided up into a number of grid points. The velocity (and also pressure and temperature) was determined by balancing the equations of fluid motion for each time step and grid point, typically milliseconds and millimeters respectively, for an engine. Atmospheric scientists also divide up the atmosphere into a grid. Because the atmosphere is so large, the grid is rather coarse. The vertical is typically divided up into ten levels, and the horizontal grid is approximately 5° by latitude and longitude (Figure 5.17).

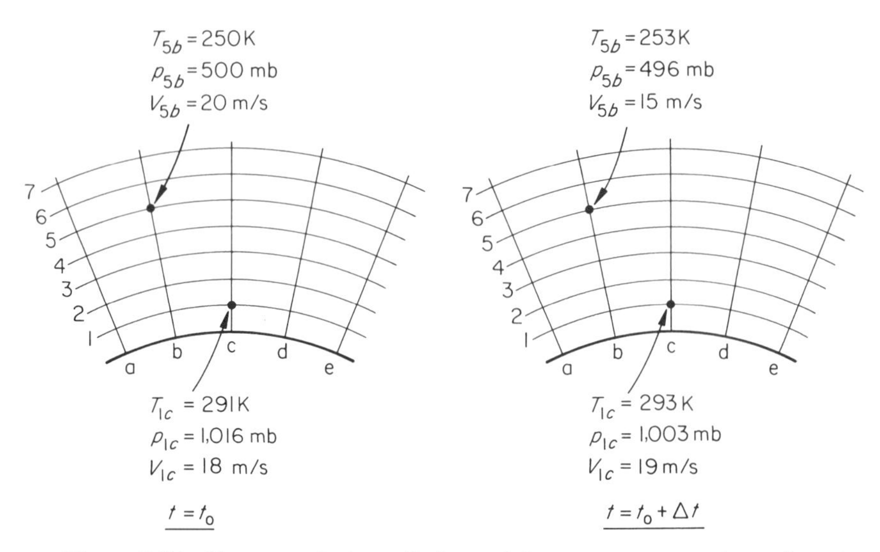

Figure 5.17 For numerical predictions of the temperature and weather, the atmosphere is divided into a grid, in a way similar to the way that the engine was divided (Figure 3.21). At each grid point the temperature pressure, wind velocity, and so forth are determined. Greenhouse gas concentrations are then increased, and these parameters are reevaluated.

At a particular time, data for the pressure, wind speed, humidity, temperature radiation intensity, and so forth for every grid point are fed into the computer. A numerical approach is then used to determine the value of those variables a time, Δt, later. As the program goes from time step to time step, the program introduces more radiation absorption into the complex set of equations, to represent increased levels of greenhouse gases such as CO_2. The programs are run to simulate many years of atmospheric evolution. Time averaging is very coarse so that daily trends are smeared out. The results of these simulations suggest significant global warming will occur if the CO_2 level of the atmosphere is doubled, but there is some controversy. Part of the problem stems from the coarse averaging, both in space and in time. Some critics feel that this averaging may be masking or smearing out some processes that play an important feedback role. There are also problems concerning the input or boundary conditions. For instance the physics of the interaction between the air and sea is very complex. There is absorption of greenhouse gases at the sea surface, and the amount of absorption is a function of how rough the sea is, of the nature of the marine life, and of the detailed chemical structure of the water near the surface. Calculating this absorption rate is difficult; it is a frontier research problem in its own right. There are similar problems concerning the interaction of the stratosphere and troposphere. And the feedback effects due to increased evaporation, anthropogenic aerosols such as sulphates, and changed cloud cover are still not fully understood.

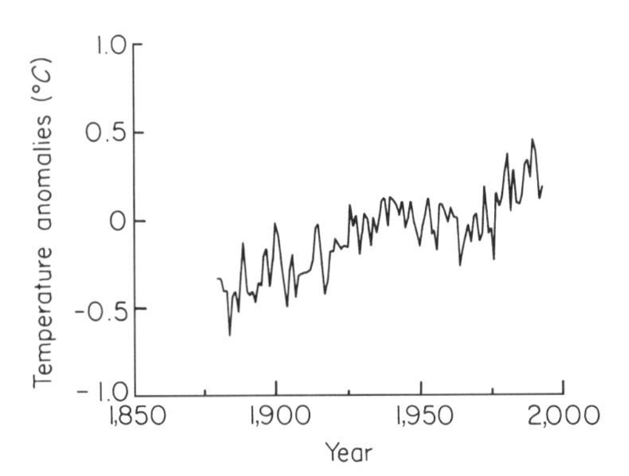

Figure 5.18 A graph of the global surface air temperature anomalies. There is a broad upward trend (although some of the dips are of comparable magnitude). (After "Trends '93," CO_2 Information Analysis Center.)

The actual global temperature record since 1900 is shown in Figure 5.18. Although it does show an upward trend, the trend is rather small and uneven and it seems to be relatively flat in the sixties and seventies when the rate of injection of CO_2 was the most rapid. On the other hand, there has been a rapid rise in the past decade. Predicting the future of this curve is a little like the visitor from another planet trying to predict the time series of Figure 5.9. Of course we do know that we are injecting CO_2 and other greenhouse gases into the atmosphere. This suggests the curve should rise even further. But as mentioned above, there may be other factors at work that we do not fully appreciate. The gentle increase over the past 100 years may be merely part of a natural long-term fluctuation in temperature. Many prominent atmospheric scientists believe that the upward trend is indeed a result of the increase in greenhouse gases and other pollutants, but proving this in a neat and tidy way is probably impossible. Idly waiting to find out the answer could be disastrous.

These considerations amplify one of the points I made in the introduction. There I suggested that often engineers have to act in the face of incomplete knowledge. Our discussion of the greenhouse effect has indicated that a full understanding may be some time in coming. Some researchers even suggest it may be too difficult to ever solve. However it is no less important because it does not have a neat solution. How do we act in the face of this incomplete knowledge? For example, what guidelines do we use in designing our new car fleets if we do not quite know what effects their exhaust is having? We will address these issues in Chapter 7.

5.3 The Atmospheric Boundary Layer

Although the atmosphere extends to some hundreds of kilometers, we have seen that most of its mass is in the troposphere, the first 10 km above the surface. In this section we focus on the layer of the troposphere closest to the ground, the atmospheric boundary layer (ABL). This is the region into which we exhaust the combustion products of our cars and power plants. Here too we grow our crops and fertilize our fields. Clearly the ABL deserves special attention.

In Section 5.1.4, I defined the ABL as the height to which frictional effects due to the no-slip condition are important. Here there is shear, because the wind speed varies from zero at the ground to the almost shearless value aloft. As you will recall (Section 3.2), shear is defined as dV/dz; that is, the change of velocity with height. The shear gives rise to turbulence and enhanced momentum transport. Furthermore, when the sunshine reaches the ground it heats it, making it warmer than the air above. So a mixture of free and forced convection transfers the heat upward, warming the ABL. We have just seen that this heat transfer plays a very important role in balancing the radiation budget.

Figure 5.19 shows an idealization of the wind and temperature profiles that we would expect to find in the ABL on a fairly typical day. The clouds usually sit at the top of the boundary layer, in an inversion region that we will discuss below. Notice the sharp temperature and velocity gradients (or wind shear) close to the ground, moderating to gentle values higher in the ABL. As we discussed in Chapters 3 and 4, this is a typical characteristic of turbulence profiles. The strong mixing tends to even out momentum and temperature differences, providing a nearly constant mean wind and temperature for most of the ABL. However, close to the ground the profiles must match the surface constraints – its temperature and the no-slip condition – and so sharp gradients must occur.

If you sit out on a deck chair on a warm, late afternoon, you will be aware of the convective activity. There is a rustling of leaves and there are gentle updraughts. You can almost feel the upward heat flux, going from the warm ground to the cooler air above. But when the sun is very low on the horizon, something very dramatic happens. The source of radiative ground heating (the sun) is turned off

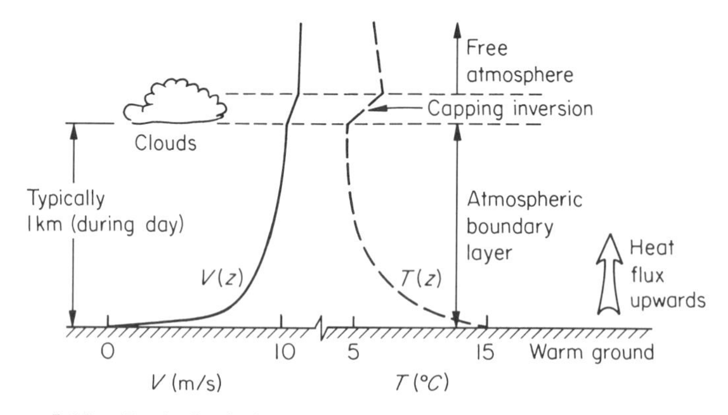

Figure 5.19 Typical wind and temperature profiles in the atmospheric boundary layer (ABL) during the day.

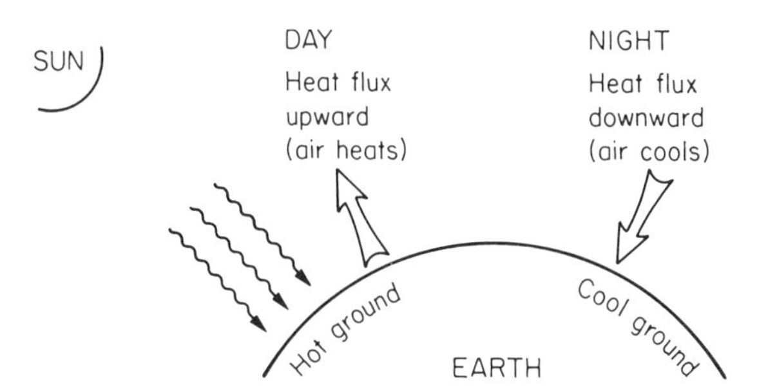

Figure 5.20 The reversal of the heat flux at dusk.

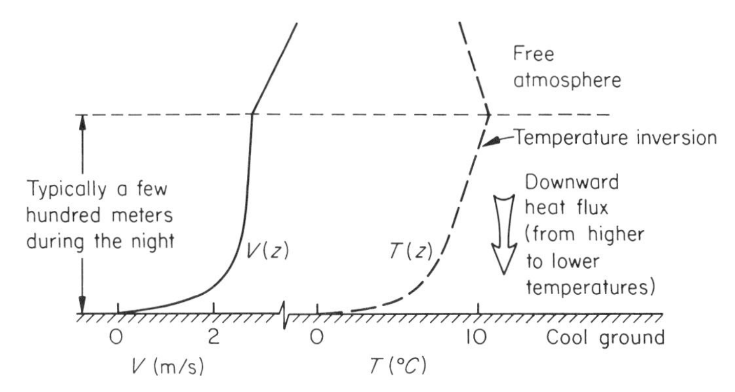

Figure 5.21 Typical wind and temperature profiles in the ABL during the evening.

(Figure 5.20). The ground cools rapidly by radiating most of the energy, stored in the first few centimeters of soil, into space. On the other hand, the air above stays relatively warm, because its heat capacity, on account of its size, is much larger than that of the first few centimeters of the ground. There results, in a matter of minutes, a complete reversal of the mean temperature gradient. It now looks like Figure 5.21. Because the temperature gradient has reversed, so too has the direction of the heat flux; it now goes downward. If you concentrate, while enjoying sitting on your deck chair, you will feel this amazing reversal. The leaves will stop moving, and there will be an eerie calm and a gentle coolness from below. This very common occurrence is best experienced inland from the sea after a warm cloudless afternoon. The strongest **temperature inversions**, as they are called, are experienced over deserts at night, but they occur everywhere.

Now we have seen, in Chapter 4, that when convection occurs, there is very efficient heat transfer in a fluid. Buoyancy promotes mixing. The temperature inversion does just the opposite; it inhibits heat transfer as well as the mixing and dispersion of smoke and pollution. After all, now the cold heavy air is near the ground, and the warmer lighter air is above it. There is little tendency for the air to move vertically under these conditions. Although beautiful to experience on a summer night, temperature inversions trap harmful pollutants. The extreme pollution of Mexico City or Los Angeles is due to temperature inversions, although their cause is usually different from the nocturnal (nighttime) mechanism just described. Clearly the temperature profile plays a very important role in the dynamics of the ABL. We will now study it in more detail.

5.3.1 Temperature Profiles in the ABL

Before we consider the interesting dynamics of what occurs when there is an inversion, we will first examine the daytime situation. As we have seen, the daytime atmosphere is in turbulent motion, and because of this it is well mixed. So apart from a very thin region close to the ground, the temperature gradient is small. However, it is not zero. In Figure 5.19 I have drawn it slightly decreasing up until the top of the ABL, and decreasing again at approximately the same rate above it.

The form of the temperature profile is determined by the way the atmosphere is mixed. Let us assume that the motion of the eddies that causes this mixing occurs sufficiently fast, so that not much heat is lost from them; that is, their motion is adiabatic. However, their volume will change as they move up and down, and so pdv work will be done. Thus, the first law of thermodynamics ($\delta q + \delta w = du$) becomes, for this adiabatic ($\delta q = 0$), compressible process:

$$-pdv = c_v dT. \qquad (5.14)$$

The atmosphere behaves as an ideal gas, and so differentiating the ideal gas law ($pv = RT$), we find

$$pdv + vdp = RdT. \qquad (5.15)$$

Substituting (5.14) into (5.15) there results

$$-c_v dT + vdp = RdT. \qquad (5.16)$$

Now the atmosphere as a whole is in hydrostatic equilibrium (Section 5.1.4), so using the hydrostatic condition (Equation 5.2) in the form, $dp = -\rho g dz$, Equation (5.16) can be written as

$$-c_v dT - gdz = RdT. \qquad (5.17)$$

(Note that $v = 1/\rho$.) We have previously shown that $c_p - c_v = R$, so Equation (5.17) becomes

$$\begin{aligned}\frac{dT}{dz} &= -\frac{g}{c_p} \\ &= \frac{-9.8}{1{,}005} = -9.75 \times 10^{-3}\ \text{K/m}. \qquad (5.18)\end{aligned}$$

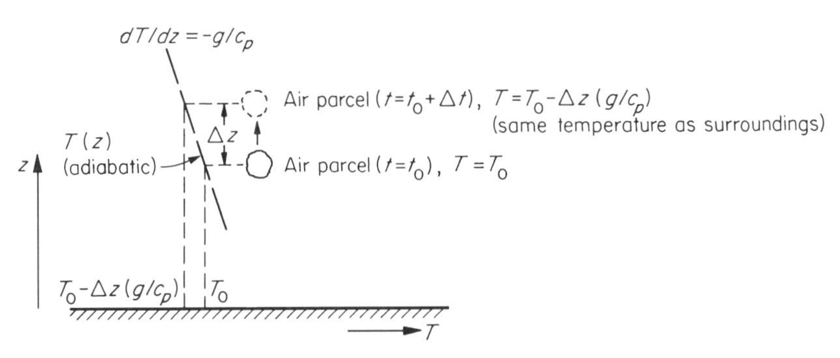

Figure 5.22 The adiabatic lapse rate.

This is a very important result. It states that if parcels or eddies of air move about adiabatically under the influence of the earth's gravity, so that the atmosphere is well mixed, there will result a decrease in temperature of nearly 10°C per kilometer. The constant, g/c_p, is called the adiabatic lapse rate.

Adiabatic lapse rates occur often, although there are exceptions. For example, in Figure 5.19, close to the ground the lapse rate is faster than adiabatic, while at the top of the ABL the gradient in temperature is positive. And during the evening there is typically a temperature inversion. We will now discuss these situations by contrasting them to the adiabatic lapse rate.

First, consider the motion of a parcel of air in an atmosphere where the temperature gradient is already adiabatic (Figure 5.22). Provided that the motion of the parcel itself is adiabatic, the temperature of the parcel of air will also decrease by 10 K/km (0.01 K/m) when moved vertically. Now, because the atmosphere itself has the same (adiabatic) lapse rate, then the parcel will move to surroundings of precisely the same temperature (Figure 5.22). Its pressure will also change according to the hydrostatic condition. So the air parcel will be in equilibrium with its surroundings; it will be neither heavier nor lighter.

Next, consider a situation (Figure 5.23) in which the air temperature increases with height rather than decreases. This may be due to the nocturnal radiation inversion that we described. Assume that somehow a parcel of air in this atmosphere is given a push from below, forcing it to rise. Now, the hydrostatic condition (Equation (5.2)) is very general, and it applies to inversion situations also. Moreover, if we assume the push is sufficiently rapid so that the parcel of air moves adiabatically, then in spite of its surroundings, the air inside the parcel will decrease in temperature at the adiabatic lapse rate. Hence, the rising parcel will become cooler than its surroundings. However, its pressure will have adjusted to them. Thus, its density will have increased relative to the surroundings, and it will fall back toward its original position (Figure 5.23). In falling it will overshoot its original position, moving adiabatically into a region that is relatively cooler, and it will be forced upward (Problem 5.10). Each time,

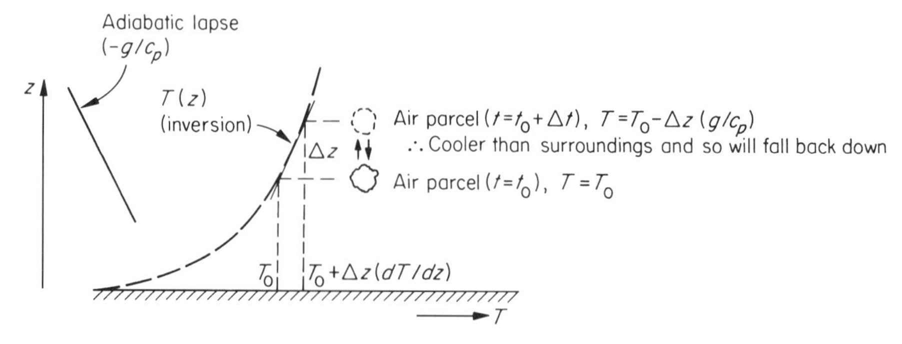

Figure 5.23 A parcel or eddy of air moving vertically in the atmosphere when there is an inversion.

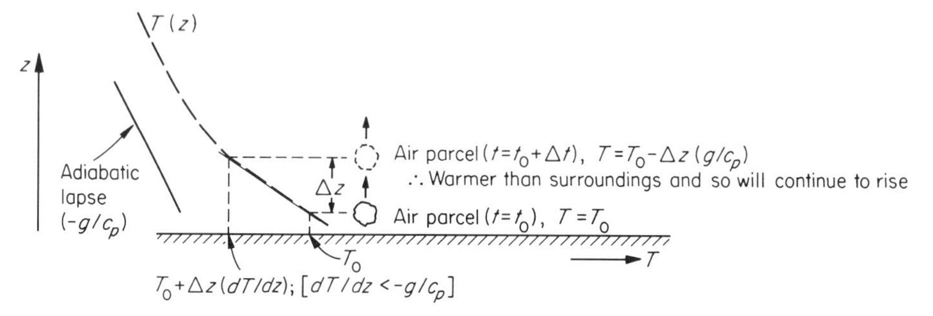

Figure 5.24 A parcel or eddy of air moving vertically in the atmosphere when the lapse rate is greater than the adiabatic lapse.

the oscillations will become smaller (because of friction and other effects) until the parcel settles back to its original position. The stronger the inversion, the smaller will be the excursions of the parcel about its origin. We say that such an atmosphere is **statically stable**, because if disturbed it returns to its original state.

Finally, consider the opposite situation: that of an atmosphere in which the lapse rate is greater than the adiabatic lapse. This occurs close to the ground in the ABL (Figure 5.19). Here if a parcel is moved adiabatically upward, it will be warmer than its surroundings and will tend to rise even further (Figure 5.24). If it is pushed down, it will be cooler, and therefore it will fall even further. For this situation the atmosphere is termed **statically unstable**.

From these considerations it is evident that the adiabatic lapse rate is a reference for the stability of atmospheric motion. If the lapse rate is greater than this (dT/dz more negative than $-g/c_p$), disturbances are amplified. This is why mixing is so rapid close to the ground on a warm day. On the other hand, if the lapse rate is less than this, such as when there is a nocturnal inversion, disturbances are damped and there is little mixing in the vertical. Note that the requirement for an inversion condition is that dT/dz be less negative than

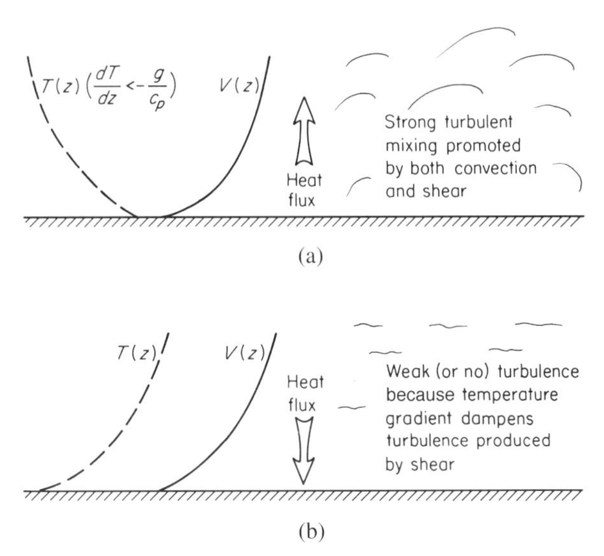

Figure 5.25 (a) Mixed (forced and free) convection. (b) Heat flux in the presence of an inversion.

$-g/c_p$. So the temperature may actually decrease with height; as long as the decrease is less than g/c_p, it will be an inversion. The inversion shown in Figure 5.23 is rather strong; here dT/dz increases with height. The essential point is that the gradient $-g/c_p$ provides the neutral case. Anything less than than this is unstable, and anything greater is stable.

5.3.2 The Richardson Number

Does the analysis of the previous section imply that whenever $dT/dz > -g/c_p$ there will be no vertical air motion at all? The answer to this is no, because so far we have not considered the wind. We have considered the static stability, but not the **dynamic stability** of the atmosphere. As we know from Chapter 3, wind shear produces turbulence. Now, if the lapse rate of the atmosphere is adiabatic or greater than adiabatic, the shear will act in concert with the convection to promote the vertical mixing. The atmosphere will be in a state of forced convection (Figure 5.25(a)). However, if there is an inversion there will be a competition between the shear that wants to promote turbulence and the density gradient that wants to dampen it (Figure 5.25(b)). More formally, we say there is a competition between the inertial force (due to the fluid motion) and the buoyancy force. The situation is described in terms of a dimensionless ratio known as the Richardson number, Ri, defined as

$$Ri \equiv \frac{[g/T_0][dT/dz + g/c_p]}{[dV/dz]^2}. \tag{5.19}$$

Here dT/dz is the temperature gradient, dV/dz is the wind shear, and T_0 is a reference temperature: the temperature where the Ri is being calculated. The numerator of this expression describes the buoyancy forces; the denominator, the inertial forces. When the temperature gradient, dT/dz, is equal to the adiabatic lapse $(-g/c_p)$ then $Ri = 0$. For this case the atmosphere is called neutral. Here mixing is promoted by wind shear only. If the temperature lapse is greater than adiabatic, that is, if dT/dz is more negative than $-g/c_p$, then Ri will be negative; if dT/dz is less than adiabatic there will be a positive Ri. Its actual magnitude will depend not only on the temperature gradient but also on the magnitude of the shear. Very strong shear will reduce the Ri to small values, even if dT/dz departs significantly from the adiabatic lapse. This makes sense because even if there is an inversion, large shear could provide sufficient turbulence to overcome its stabilizing effects. A combination of experimental and theoretical work has shown that if the Ri is greater than 0.25, turbulence will be damped, but for values less than this there will be turbulence. Thus, even if there is an inversion, if there is sufficient wind shear then it is still possible to have turbulent mixing.

We will consider some examples (Figure 5.26). First, assume that very close to the surface the lapse rate is greater than adiabatic (that is, it is more negative than -0.0097 K/m), say -0.012 K/m or -12 K/km. Assume also that there is much wind shear (Figures 5.19 and 5.25 show that this is generally the case close to the surface). Let us assume the wind speed changes from 1 m/s to 2 m/s in 5 m. Thus $dV/dz \sim 1/5\ \text{s}^{-1}$. If $T_0 = 300$ K, and because $g/c_p = 0.0097$ K/m, then from Equation (5.19), the Ri is -1.9×10^{-3}. This is close to zero, indicating that the shear is controlling the dynamics of the mixing. On the other hand, consider an inversion with $dT/dz = 0.1$ K/m, a fairly typical nocturnal value. Again, assume the same shear, of $0.2\ \text{s}^{-1}$, as before. Here $Ri = 0.09$, again a small value, but this time positive. Here, too, although there is an inversion, the shear is strong enough to promote turbulence because the Ri is less than 0.25. Now, assume the shear drops from $0.2\ \text{s}^{-1}$ to $0.02\ \text{s}^{-1}$. The Ri will increase 100-fold

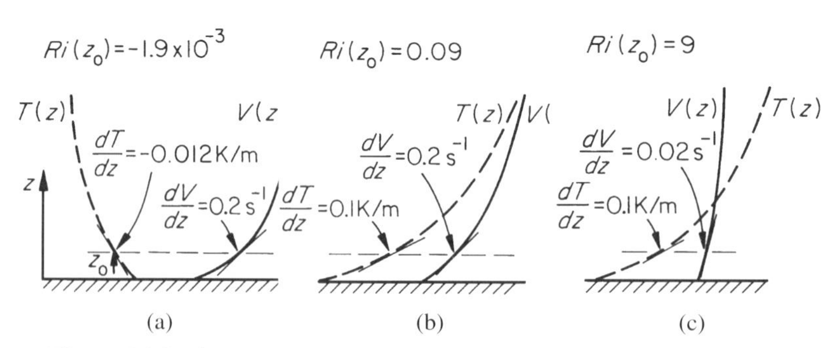

Figure 5.26 Some examples of different Richardson numbers.

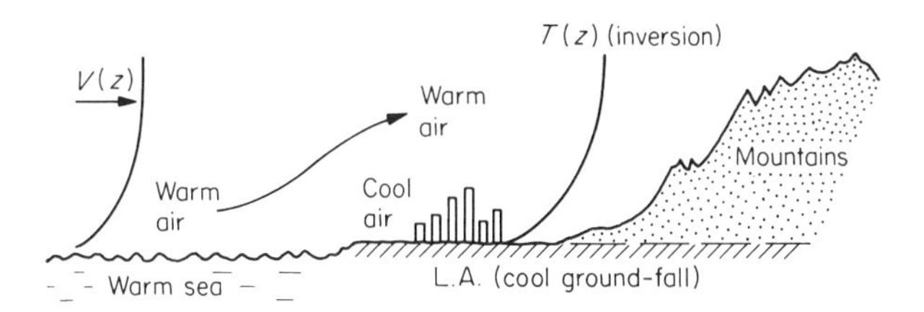

Figure 5.27 A typical inversion caused by warm air overriding cool air. Such inversions can last for days.

to a value of 9. Now the buoyancy forces of the inversion are too strong for the shear to overcome, and turbulence ceases.

High values of Ri frequently occur at night when there is a radiation inversion. However, they may also occur at any time of day if for some reason warm air overrides cooler air below. This frequently occurs in coastal regions, particularly if there are surrounding mountains. Consider, for example, Los Angeles (Figure 5.27). In the fall, the sea temperature may be warmer than that of the land. Air blowing over the sea will be warmed, but the air above the land will be cool. When the warm air from the sea reaches the coast, it will override the cool air and an inversion will occur. The mountains will block the air motion, so the shear will be small. A high Ri will result.

It should be apparent that temperature inversions or, to be more accurate, high Richardson numbers promote severe pollution because vertical mixing is suppressed. In the early fifties thousands of people died in London because of an inversion that lasted for days. The main source of heating was from coal fires. Millions of chimneys were discharging SO_2 and other dangerous gases into the still air. It did not mix and so it accumulated, causing asphyxiation. Residents of Athens, Bangkok, and Mexico City also experience severe pollution episodes, and these cause chronic illness. Here sources of pollution are automobile and power plant exhaust and leaks from bottled gas used for cooking.

Figure 5.28 shows a smokestack in a convective, normal daytime situation and in an inversion. For the convective case the smoke disperses and dilutes quite rapidly. For the inversion, although the smoke will rise a little because of its initial buoyancy (it will generally be warmer than the surrounding air), it becomes trapped when its temperature reaches that of its surroundings. Because there is no vertical mixing, it sits there. Of course the situation is generally more complex than this, because there is usually some wind shear and the surrounding topography may allow for some drainage. We will discuss the pollution from smokestacks and automobile exhausts again in Chapter 7.

We have concentrated on the thermal characteristics of the ABL because of their role in pollution episodes. But just as there are temperature gradients,

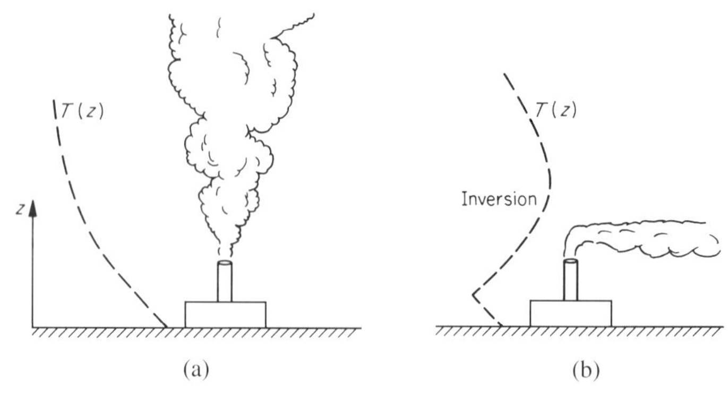

Figure 5.28 A smokestack in a convective atmosphere (negative Ri) and in a stably stratified atmosphere (positive Ri).

there are humidity gradients, as well as gradients of other species such as CO_2 and methane. Above a water surface, the humidity generally decreases, and thus there is an upward flux of water vapor. An understanding of its transport characteristics is important for water conservation as well as weather prediction, both long and short term. The presence of water vapor also alters the "dry" lapse rate, reducing it from −9.7 K/km to around −6 K/km, although this depends on the amount of humidity and the average temperature. These and other aspects of the ABL are dealt with in texts on meteorology and atmospheric physics.

5.4 Discussion

In this brief introduction to the atmosphere we have focused on two aspects: the greenhouse effect and the form of the temperature profiles in the first few hundred meters of the atmospheric boundary layer. An understanding of these is required in order to determine the effect of the engine on the atmosphere. Yet there are many other aspects of the atmosphere that are equally important, and some of these are also relevant to the engine–atmosphere interaction. For example, we have not dealt with the large-scale motion of the atmosphere, except in passing. To properly understand the overall dynamics of the atmosphere entails a detailed study of the equations of motion. (These are described in the appendix.) The formation of clouds and raindrops, the flow of air over mountains, and the interaction of the air with the sea are other topics that are central to understanding the atmosphere. Some of you will be tempted to study these topics later. As I have already mentioned, an engineering degree is excellent preparation.

I also focused (in Section 5.1) on another, more general characteristic of the atmosphere. This was the interaction of the vast number of time, and hence length, scales. The scale of the motion varies from large-scale weather systems of thousands of kilometers down to gusts of the order of centimeters. Their dynamical interaction presents formidable problems to forecasting, both for the

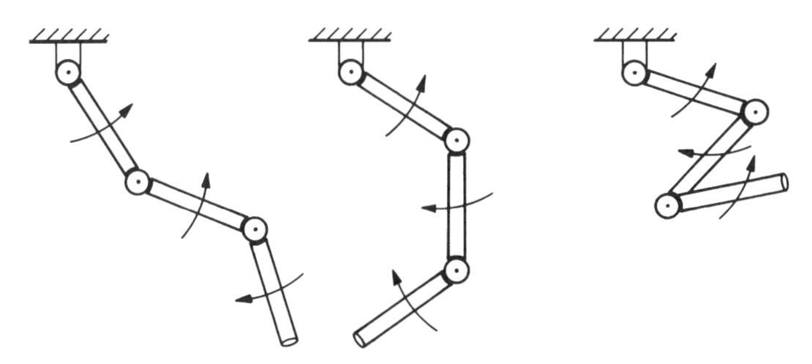

Figure 5.29 A chaotic pendulum.

short and very long term. There are results from chaos theory that show that when there is a complex interaction of various modes of a system, prediction can become impossible, even if at some moment the whole system is completely described. If we return to our pendulum example again, but now with three arms rather than one, it can be shown that it can be excited in such a way that its motion will never repeat itself (Figure 5.29). Such complex motion is unpredictable. The three modes of the pendulum should be compared with the millions of modes of the atmosphere. Things, fortunately, are not quite as bad as this. The constant deterministic forcing due to the earth's rotation, and nocturnal and seasonal variations, give some order to the system. Nevertheless, it remains an open question whether we will ever be able to properly determine next year's average temperature, or how much warming will occur if we load the atmosphere with more CO_2 or CH_4. These are some of the most important, and yet perplexing, questions of our time.

5.5 Problems

5.1 The surface temperature and radius of the sun are 5,770 K and 7.0×10^8 m respectively.
 (a) Verify that the average power at the edge of the earth's atmosphere is 343 W/m^2.
 (b) On a clear night under a full moon one can almost read a newspaper. This is reflected sunlight. The radius of the moon is 1.7×10^3 km, and it is 3.8×10^5 km from the earth. Assuming that 50 percent of the incident sunlight is reflected from the moon, estimate the power of the moonlight per unit area, distributed over a plane at the earth–moon distance. Assume the moon is a disk and the sunlight is diffused into a hemisphere by its rough surface.
 (c) Check whether your answer to (b) is reasonable by calculating the visible light power per unit area 2 m from a 60-W lightbulb. Note: incandescent lightbulbs are very inefficient. Only 5 percent of the power is in the visible part of the spectrum.

5.2 Measurements of the fluctuating velocity in the atmospheric boundary layer (ABL) show that the large eddies, which carry most of the turbulent energy, have a length scale, ℓ, of order 1,000 m and their fluctuating velocity, w, is about 1.3 m/s. (The rapid variation in the fluctuating velocity of Figure 5.7 is approximately this value. It is due to those large eddies.) In Chapter 3 we discussed how the energy of the turbulence is dissipated into internal energy by means of a cascade and that ϵ, the rate at which turbulence energy (per unit mass) is produced, must be the same as that dissipated. ϵ (W/kg or m^2/s^3), is of the order w^3/ℓ.

(a) Determine ϵ for the ABL (the first km above the ground) and then determine the total rate of kinetic energy dissipation (in kW) for the ABL. Take $\rho = 1.2\,\text{kg/m}^3$. About 3/5 of the total atmospheric kinetic energy is dissipated in the ABL (where the velocity shear is large). Determine the total rate of energy dissipation for the whole atmosphere. Your answer should be consistent with the value given in 5.1.1 for the rate at which the incident energy from the sun is converted into atmospheric motion. Why?

(b) Should the conversion of the kinetic energy of motion into internal energy cause the atmosphere to warm? (Assume that the turbulent energy of Part (a) was dissipated in an adiabatic container and determine by how much the temperature would rise in, say, 24 hr.)

5.3 In the previous problem we required an estimate of the typical fluctuation velocity, w, in the atmospheric boundary layer. These fluctuations are usually caused by the mean wind shear, dV/dz (see Section 5.3). The wind shear creates eddies whose size, ℓ, grows with height above the ground. Consider a typical large eddy of size 1,000 m and a typical wind shear of 1 m/s per km. Use a dimensional analysis approach to write a formula for the characteristic velocity of the eddy, w, in terms of dV/dz and the length scale. Determine the fluctuation velocity and show that it is consistent with the estimate used in Problem 5.2.

5.4 In cold climates (such as where I live in upstate New York) the air becomes very dry and home humidifiers are used in winter. They often consist of a reservoir of a few gallons of water with a fan blowing over its surface. I notice that my humidifier typically evaporates about 2 quarts ($1.9 \times 10^{-3}\,\text{m}^3$) in 8 hr. Determine the amount of energy required to do this.

The fan on the humidifier is rated at 100 W. Is this providing the required amount of energy to evaporate the water? What is the fan really doing? (Hint: Would any water evaporate without the fan?)

5.5 (a) Consider a moist atmosphere, consisting of dry air and water vapor. Show that the mean molecular weight, $\bar{M}$, of the moist air is

$$\bar{M} = [(m_d/M_d + m_v/M_v)/(m_d + m_v)]^{-1},$$

where m is the mass of each constituent, M is its molecular weight, and the subscripts d and v refer to dry air and water vapor respectively.

(b) Show that if the partial pressure of the water vapor, e, is much less than that of the mixture, p, the **mixing ratio**, m_v/m_d, is given by the expression

$$q_v \equiv \frac{m_v}{m_d} \approx \frac{M_v}{M_d}\frac{e}{p}.$$

(c) The equation of state for moist air is $pv = R_m T$, where p and v are the pressure and specific volume of the mixture respectively, and R_m is the gas constant of the mixture. Determine R_m in terms of q_v, M_v, M_d, and R_u, the universal gas constant.

(d) Modify the way you have written your answer to part (c) to be in the form

$$pv = \frac{R_u}{M_d}T^*,$$

where again p and v are the pressure and specific volume of the mixture, R_u is the universal gas constant, and T^* is a modified temperature, called the **virtual temperature**. It is the temperature that dry air would have if its pressure and specific volume were equal to those of a given sample of moist air. Assuming the maximum ratio of moisture to dry air, m_v/m_d, is 4×10^{-2} (i.e., 40 g of water vapor for every kg of dry air), what is the maximum difference between T^* and T? Is T^* greater or less than T? Why?

5.6 Assume that in the troposphere there is a region in which the temperature decreases linearly with height:

$$T = T_0 - \gamma z.$$

Here T_0 is the temperature at the bottom of the layer, and γ is the lapse rate in temperature (K/m).

For a linear temperature decrease γ must be constant. Using the perfect gas law and the hydrostatic equation, determine the distribution of pressure with height for this case. Let the pressure at the bottom of the layer be p_0.

For a 2-km layer, with $T_0 = 300$ K, $p_0 = 10^5$ N/m^2, and $\gamma = 10$ K/km, determine the change in pressure across the layer. Compare this with the pressure change for this height interval assuming the exponential pressure decrease (isothermal atmosphere) discussed in Section 5.1.4.

5.7 (a) Determine the total heat capacity (J/K) of the atmosphere and of the oceans. For water, $c = 4{,}200$ J/(kg K). The values of other parameters are given in Section 5.1.

(b) Determine the heat capacity of a 1-m^2 column of the atmosphere. Use the characteristic scale height of the atmosphere.

(c) Consider a small change in temperature of the column of part (b). Assume that the column will equilibrate to its original temperature by means of radiation only. Determine the characteristic time, τ, that it takes. (Hint: because radiation is involved, τ must be a function of σ and T. It also will depend on the mass of the column (ρH) and its specific heat. You may use dimensional analysis to find the order of magnitude of τ.)

5.8 Assume a simple atmosphere in which CO_2 is the only gas that plays a role in the greenhouse effect. In the text (Section 5.2) we showed that a doubling of CO_2 would cause a 1 K rise in the temperature of the atmosphere. Using Equation (5.13) determine how long the atmosphere will take to reach 95 percent of its final temperature. Draw a graph of the process. As in the text, take $\lambda = 4.8\ \mathrm{W/(m^2\ K)}$ and assume the doubling of CO_2 will increase the absorbed power by $4\ \mathrm{W/m^2}$. Is it a coincidence that your answer is of the same order as part (c) of Problem 5.7?

5.9 There is much discussion about a rise in sea level due to global warming. Even a small amount of melting of the polar ice caps could lead to a large rise in sea level. Yet scientists are still unsure of whether global warming will cause ice cap melting. For example, increased temperatures could cause increased evaporation and thus more precipitation on our ice caps resulting in yet more ice. However, it is certain that if the atmosphere warms so must the oceans. If the mean density of sea water is $1.037 \times 10^3\ \mathrm{kg/m^3}$ and it decreases by one part in one thousand per 1K rise in temperature, determine the rise in sea level for a 5 K warming of the top 100 m of the oceans.

5.10 (a) Under temperature inversion conditions, a small parcel of air will oscillate up and down if displaced vertically from its original position (see Figure 5.23 and the related discussion). The frequency of oscillation will be a function of the temperature gradient, the mean temperature, and g, the acceleration due to gravity. Determine, using dimensional analysis, the dependence of f on these three parameters. For an inversion condition of a few degrees K per 100 m, would the oscillation period be seconds, minutes, or hours?

(b) What is the magnitude and sign of the Richardson number for free convection?

(c) In practice a value of $Ri \sim -1$ implies that free convection is very significant. Consider a wind shear of 1 m/s per hundred meters. Assume the wind is increasing linearly, from zero at the ground. For this situation what value of dT/dz would yield a value of $Ri = -1$. How does this compare to the adiabatic lapse rate?

5.11 (a) In Problem 4.7 the flow is stably stratified. Assuming the depth of the boundary layer, ℓ, (Figure 4.19) is 10 m, determine the bulk Richardson number, R_{i_B}, for this flow: $Ri_B \equiv (g/T_0)[\Delta T/\ell + g/c_p]/(\Delta U/\ell)^2$.

Here ΔT and ΔU are the respective changes in temperature and velocity over the boundary layer depth. Does your answer suggest that the stable stratification is playing a major role in the suppression of the heat transfer?

(b) Now determine the Nusselt and bulk Richardson numbers for the same ice field but with a cool breeze of $-5°$C. (The wind speed, ice temperature, and boundary layer depth remain the same.) In which direction is the heat flux? Is there a practical cooling problem now? (Does it matter if the ice equilibrates to the air temperature of $-5°$C? If it is required to stay at $-1°$C, how much power is needed?)

Symbols

A	area	m^2
c_p	specific heat at a constant pressure	J/(kg K)
c_p'	specific heat (constant p) per unit area of the atmosphere	J/(m^2 K)
c_v	specific heat at a constant volume	J/(kg K)
E_b	black-body radiation power	W/m^2
f	frquency	Hz
g	acceleration due to gravity	m/s^2
H	scale height	m
ℓ	turbulence length scale	m
m	mass	kg
p	pressure	Pa
R	gas constant	J/(kg K)
Ri	Richardson number	
R_{net}	net solar power per unit area into earth's climate system	W/m^2
S_0	solar power per unit area at mean sun–earth distance	W/m^2
T	temperature	K
T_e	radiating temperature of the earth–atmosphere system	K
T_s	average temperature of the earth's surface	K
t	time	s
u	internal energy per unit mass	J/kg
V	velocity	m/s
v	specific volume	m^3/kg
w	turbulence velocity scale	m/s
z	height	m
α	albedo	

$\Delta \dot{Q}$	change of rate of heat interaction per unit area	W/m^2
λ	feedback factor	$W/(m^2\ K)$
ρ	density	kg/m^3
ρ_0	reference density	kg/m^3
σ	Stefan–Boltzmann constant	$W/(m^2\ K^4)$
τ	turbulence time scale	s

6

Energy Sources

In Chapter 2, on thermodynamics, we studied heat engine cycles and their efficiency. Recognizing that the stuff inside engines is usually a gas or liquid, we studied fluid dynamics in Chapter 3. In Chapter 4 we examined the way heat is transferred to and from the engine, and within it. We also saw the intimate connection between heat transfer, fluid dynamics, and thermodynamics. To complete our story of the engine, we must turn to the method by which we produce the high-temperature sources that run it.

In a literal sense, all of our energy is solar. Hydrocarbon fuels that run nearly all of our automobiles and most of our power stations are formed from organic matter, which is the result of photosynthesis, the process by which sunlight, CO_2, and water form plants and trees. Hydroelectric power and wind power are also intimately connected with the sun; for the former the sun evaporates the water that then condenses to fill reservoirs that provide the water to drive turbines, and for the latter the sun provides the source of energy for the winds. Even nuclear power is connected to the sun, because the very chemical structure of our minerals, from which nuclear fuels are made, was formed from the primeval sun.

At present, hydrocarbon fuels are by far the most important sources of energy both in terms of the power they produce and their effect on the environment, and for this reason they are a central focus of this book (Figure 6.1). Nuclear energy is likely to assume a greater role in the future, particularly for the smaller well-developed countries that wish not to depend on other countries for hydrocarbon fuels. Even now, in France, most of the energy used in industry and in the home comes from electricity generated using nuclear reactors as the high-temperature heat source. On a global scale, hydroelectric, wind, geothermal, and direct solar energy presently play a peripheral role, but they almost certainly will increase in importance as their cost decreases and as the environmental dangers of other energy sources become more pronounced.

Figure 6.1 A smokestack from a coal-burning power plant. This photo was taken on a cold winter morning. The water vapor has condensed, making the plume visible. Apart from water the other main product of combustion is CO_2. There are also traces of NO, NO_2, O_3, SO_2, and other constituents that form pollution. These will be discussed in Chapter 7. (Courtesy of Mr. E.P. Jordan, Cornell University.)

In a strict sense, I should not use the expression "energy source." After all, the first law of thermodynamics states that energy (or mass-energy in the case of nuclear reactions) must be conserved. So literally speaking we cannot have an energy source; we can only convert one kind of energy to another: internal energy to kinetic energy of motion and so on. Nevertheless, it is common parlance to use the expression "energy (or power) source," rather than "energy conversion mechanism," and I will keep to this convention here. It is the purpose of this chapter to summarize the main sources of energy available to us. The physics and chemistry are complex, particularly for combustion processes, which paradoxically are still less well understood than nuclear processes. Here I will provide some of the basics. The details must wait for further courses.

But first, a note of caution. When you hear people discussing energy, you will inevitably hear an argument. One person likes nuclear energy; the other thinks it is deadly and wants solar energy. Another thinks that the global warming issue is a plot by academic extremists and advocates bigger gasoline auto engines and cars with better acceleration. Rarely will you hear these people argue over the merits of forced or free convection or the validity of the second law of thermodynamics! When you read this section you will find that I, too, after summarizing the mechanisms of each method of energy production, will outline their pros and cons.

Why does energy production elicit great emotion, from technical and non-technical people alike? At least part of the reason, I believe, is that energy is a metaphor for the technological society in general. Ecologically minded people believe solar energy is better for the environment. It is cleaner, and there are no black smokestacks. Others, who wish to see industrialization, will advocate the use of any energy source so long as it is easily available. They argue that environmental concerns will sort themselves out when they become really oppressive. Highly technical people might favor rapid research and development in nuclear fusion because it represents an exciting challenge. All of these people are channeling a part of their world view into the discussion. Gone is the supposedly dispassionate, analytical tone of the scientist or technologist.

6.1 The Energy in Chemical Bonds: Hydrocarbon and Hydrogen Fuels

Heat has been produced by burning fossil fuels from prehistoric times. A simplified equation for the burning (combustion) of wood or lumber is

$$CH_2O + (O_2 + 3.76N_2) \rightarrow H_2O + CO_2 + 3.76N_2. \qquad (6.1)$$

The molecular structure of wood is in fact much more complex than CH_2O, which may be considered as an elementary molecule from which the complex carbohydrates (starch and cellulose) are formed. Notice here, as in Chapter 1 (Equation (1.2)), I have included nitrogen, the major constituent of the atmosphere. (There are 3.76 moles of N_2 for every mole of O_2 in the atmosphere.) It plays no role in the above reaction: the energy released is due to the conversion of CH_2O into H_2O and CO_2. We will show in Chapter 7, however, that the nitrogen plays a major role in creating pollution by combining, in small amounts, with the oxygen.

Approximately 10^7 J of energy is released by the combustion of 1 kg of wood. This energy is due to the breaking of the interatomic or chemical bonds of the CH_2O molecule. Because there are 6.02×10^{26} atoms/(kmol) – Avogadro's number – and the molecular weight of CH_2O is 30, this means that $30 \times 10^7/(6.02 \times 10^{26}) = 5.0 \times 10^{-19}$ J of energy is released for every wood molecule burnt. This is equivalent to the kinetic energy of an undernourished baby bug with a mass of 1 mg moving at 3.6 mm/hr! It is the extraordinary number of molecules in a kilogram of wood that provides the vast amount of energy released, not the energy per molecule. When talking about small amounts of energy, we often use the electron-volt (eV) as the unit. It is the energy an electron would gain in falling through a potential difference of 1 V. 1 eV $= 1.6 \times 10^{-19}$ J. Thus one CH_2O molecule releases approximately 3 eV. The energy released in most combustion processes is of the order of 1 to 10 eV per molecule. We will contrast this to nuclear energy in the next section.

Special relativity theory tells us that if energy is created, then there must be a depletion of mass according to the equation $E = mc^2$. (Here c is the velocity of light (3×10^8 m/s).) So burning 1 kg of wood and thereby producing 10^7 J would result in a mass change of $E/c^2 = 1.11 \times 10^{-10}$ kg. This is unmeasurable, and this is why we say mass is conserved in chemical reactions. It is not strictly not true, and more formally we should talk of mass-energy conservation.

Let us look at the reverse of Equation (6.1):

$$H_2O + CO_2 \rightarrow CH_2O + O_2. \tag{6.2}$$

(Here for simplicity I have neglected the N_2.) The energy for this reaction (it absorbs energy, rather than releasing it) comes from sunlight. This reaction is a crude model of photosynthesis. As you may know from biology, the details are very complex indeed. To make 1 kg of biomass from photosynthesis may take years, whereas to burn 1 kg (Equation (6.1)) takes only a few minutes.

When we look at the more complex hydrocarbons (such as octane, C_8H_{18}) that are the main constituents of our fossil fuels, their production, by means of biomass decay and by subsequent high-pressure processes deep below the earth's surface, takes hundreds of millions of years. But again, as for wood, their combustion takes only a few minutes. Thus, although the equations for fossil fuel combustion are reversible, unfortunately the reaction times are different by factors ranging from 10^6 to 10^{13}. This is why oil is so valuable; its replacement time is on a geological rather than on the human time scale.

Typical fossil fuels produce energy in the range 10^7 J/kg (blast furnace gas) to 5×10^7 J/kg (octane and methane). As we have seen in Chapter 1 (see for example Problem 1.5), this is a big kick from our kilogram, or a big bang for our buck! Indeed it is more than the energy that is produced from 1 kg of TNT explosive, although the reaction time is slower. An order of magnitude less energy per kilogram than this would make fossil fuels impractical, at least for vehicles. Imagine a subcompact car with a 100-gallon tank.

The basic chemical reaction equations for fossil fuels are all of similar form to Equation (6.1), be they gasoline, natural gas, or coal. As we emphasized in Chapter 1, they all produce CO_2 and H_2O (Figure 6.1). However, the combustion process is never perfect, producing only CO_2 and H_2O. For example, Problem 1.3 shows a reaction of propane with a 20 percent deficit of air (a so-called rich fuel-to-air ratio). Here carbon monoxide, a highly toxic gas, was also formed. Gasoline, too, produces CO in small amounts when burned rich. We emphasized in Section 1.2 that very small numbers multiplied by extremely large numbers produce very large ones. So, even a few tens of grams of CO (per kilogram of fuel burned) will result in tons of CO in the atmosphere in a region such as Los Angeles where there are millions of cars.

The actual combustion process is much more complicated than Equation (6.1) or Equation (1.2) for octane combustion. There are a number of intermediate stages in the reactions that are dependent on such factors as the amount of air, the temperature of the reaction, the levels of turbulence, and the presence of other trace constituents. Figure 6.2 shows some of the reactions that occur in methane combustion. In fact there are approximately 150 reactions, each occuring at a different rate and producing (or absorbing) different amounts of heat. There is presently a great deal of research on this subject, which is known as chemical kinetics. This research is partly aimed at providing more efficient combustion, but the principal objective is to reduce the very small amounts of harmful products that are always present in combustion processes. For example, as I mentioned, the N_2 in the air combines with oxygen in part of the reaction in an IC engine to produce NO and NO_2 (see Chapter 7). Both of these gases are toxic, and they also play a major role in photochemical smog formation. Their concentration in

$+H$
$+H = CH_3 + H_2$
$CH_4 + OH = CH_3 + H_2O$
$CH_4 + O = CH_3 + OH$
$CH_4 + HO_2 = CH_3 + H_2O_2$
$CH_3 + OH = CH_2O + H_2$
$CH_3 + OH = CH_3O + H$
$CH_3 + O = CH_2O + H$
$CH_3 + O_2 = CH_3O + O$
$CH_3 + O_2 = CH_2O + OH$
$CH_3 + HO_2 = CH_3O + OH$
$CH_3 + HO_2 = CH_4 + O_2$
$C_2H_6 = CH_3 + CH_3$
$C_2H_4 + H_2 = CH_3 + CH_3$
$C_2H_5 + H = CH_3 + CH_3$
$CH_3O + M = CH_2O + H + M$
$CH_3O + O_2 = CH_2O + HO_2$
$CH_2O + M = HCO + H + M$
$CH_2O + O = HCO + OH$
$+H = HCO + H_2$
$CH_3 = HCO + CH_4$
$= HCO + H_2O_2$
$HCO + H_2O$
$CO + M$

$CH_2 + O = CH + OH$
$CH_2 + H = CH + H_2$
$CH_2 + O_2 = HCO + OH$
$CH_2 + OH = CH + H_2O$
$CH_2 + O = CO + H + H$
$CH_2 + O_2 = CO + H + OH$
$CH_2 + O_2 = CO + H_2O$
$CH_2 + O_2 = CO_2 + H + H$
$CH + O_2 = CO + OH$
$CH + O_2 = HCO + O$
$CH + O = CO + H$
$C_2H_6 + CH_3 = C_2H_5 + CH_4$
$C_2H_6 + H = C_2H_5 + H_2$
$C_2H_6 + OH = C_2H_5 + H_2O$
$C_2H_6 + O = C_2H_5 + OH$
$C_2H_5 = C_2H_4 + H$
$C_2H_5 + O_2 = C_2H_4 + HO_2$
$C_2H_4 + M = C_2H_2 + H_2 + M$
$C_2H_4 + \dot{M} = C_2H_3 + H + M$
$C_2H_4 + OH = CH_3 + CH_2O$
$C_2H_4 + O = CH_3 + HCO$
$C_2H_4 + H = C_2H_3 + H_2$
$C_2H_4 + OH = O$
$C_2H_4 + O =$
$C_2H_3 + M$

$HO_2 +$
$HO_2 + H$
$HO_2 + OH =$
$HO_2 + H_2 = H_2O$
$HO_2 + O = O_2 + OH$
$H_2O_2 + M = OH + OH + M$
$H_2O_2 + O_2 = HO_2 + HO_2$
$H_2O_2 + OH = H_2O + HO_2$
$H_2O_2 + H = H_2O + OH$
$O_2 + M = O + O + M$
$N_2 + M = N + N + M$
$NO + M = N + O + M$
$N + O_2 = N$
$O + N_2 =$

Figure 6.2 Some of the reactions that occur for the combustion of methane in air. There are nearly 150 reactions, each occuring at different rates and yielding (or absorbing) different amounts of energy. Notice the presence of O, H, OH, and NO molecules. Their significance will be discussed in Chapter 7. (After Kravchik et. al., 1996.)

the exhaust is not high, a few grams for every kilometer traveled in an average car. But this means that as for CO production, the millions of cars in the Los Angeles or Mexico City area will produce tons of NO and NO_2 per day, and this is enough, because pollution concentrations need not be high (they are usually parts per billion of air by volume), to cause severe discomfort and harm.

I will conclude this section with one of the most simple of all chemical reactions, that of hydrogen and oxygen:

$$2H_2 + O_2 \rightarrow 2H_2O. \tag{6.3}$$

This yields approximately 1.4×10^8 J/kg, nearly three times the energy of gasoline per kilogram and ten times that of wood. Notice that unlike fossil fuels, which always produce CO_2 (a greenhouse gas), hydrogen, in principle, produces no harmful products of combustion. Unfortunately, there are only minute amounts of hydrogen in the atmosphere or in the earth. Production of H_2 in the laboratory is relatively straightforward (by means of electrolysis) and, with its high energy yield and its lack of greenhouse gas products, the benefits of using wind, solar, or nuclear power to run large electrolysis plants have been widely discussed. The hydrogen could then be used as pollutionless fuel for automobiles (see Section 6.3.2).

6.2 The Energy in Atomic Bonds: Nuclear Energy

When excited properly (by means of raising its temperature) a hydrocarbon fuel undergoes a chemical reaction in which the bonding forces between the atoms are rearranged. We have just shown that the energy released is of the order of 10 eV per molecule. Within the atom is the nucleus, made from neutrons and protons. The nuclear bonding forces between these atomic constituents are much stronger than the chemical bonding forces between the atoms. When excited properly, for instance when a nucleus of uranium 235 captures a neutron, the nucleus undergoes a reaction in which the atomic bonds between the neutrons and protons are broken. Here the energy released per nucleus is typically hundreds of millions of electron volts. The neutron and proton can be further divided, and the energy involved is measured in tens of billions of electron volts. Large particle accelerators are used to study the binding forces in neutrons and protons, but here our focus will be on atomic rather than subatomic bonds.

Nuclear energy is released in two ways: by means of **fission**, whereby a large atom such as uranium 235 breaks into two fragments, or by means of **fusion**, whereby two small atoms (low atomic mass) such as two hydrogen atoms join together to make a more complex atom, in this case helium. Nuclear fission can be controlled and is the only way commercial nuclear energy is produced at present. Nuclear fusion, the source of energy for the sun, stars, and bombs, remains a promising source of commercial energy, but in spite of a great deal of research, it has eluded the engineers' attempt to harness it.

A typical fission reaction is as follows:

$$n +{}^{235}_{92}\mathrm{U}_{143} \rightarrow {}^{144}_{56}\mathrm{Ba}_{88} + {}^{89}_{36}\mathrm{Kr}_{53} + 3n + 177\,\mathrm{MeV}. \tag{6.4}$$

In this reaction uranium 235, consisting of 92 protons and 143 neutrons (left and right subscripts), absorbs one neutron, n, and fissions (splits) into two lighter nuclei, barium 144 and krypton 89. Three extra neutrons are also produced, and these, when their energy is properly moderated, go on to fission further uranium atoms, thereby causing a chain reaction. Gamma rays (very short electromagnetic radiation; see Figure 4.15) plus 177×10^6 eV of energy per atom of uranium are also produced. Notice that the total number of atomic particles is conserved ($1 + 235 = 144 + 89 + 3$). The enormous amount of energy released (compared with approximately 10 eV for a hydrocarbon molecule) is due to the release of the atomic binding energy. One kilogram of ^{235}U contains $6.02 \times 10^{26}/235 = 2.56 \times 10^{24}$ atoms. Thus 1 kg of uranium would produce $177 \times 10^6 \times 2.56 \times 10^{24} = 4.53 \times 10^{32}$ eV $= 7.25 \times 10^{13}$ J if all of the atoms were to undergo fission. From the mass–energy relationship, $E = mc^2$, this is equivalent to a mass of 0.81 g. Thus, less than one thousandth of the mass of uranium is converted into energy even if there is complete fission. Although small, it is measurable, whereas we found that the mass loss in a chemical reaction is not.

The enormous amount of energy per atom, compared to that per molecule in fossil fuels, makes nuclear fission an extremely attractive energy source. Unfortunately, there are problems here too. First, because of the shielding needed for the radiation, it is impractical to build nuclear-powered automobiles. Nuclear energy can be used only for commercial power generation and for very large ships. Second, the products of the nuclear reaction are radioactive. Some of these products have very short half-lives, the time over which one-half the radioactive nuclei in a given sample will decay. On the other hand, others, such as strontium 90 and cesium 135, have half-lives of 29 and 10^7 years respectively. The radiation produced in the decay process is extremely harmful, and so the waste products of the reactor must be stored surrounded by thick shielding. Fortunately, the total volume of the waste products is relatively small, because, as we have shown, nuclear reactions produce a great amount of energy per kilogram of fuel. The total waste products per year from all the commercial reactors operating in the United States could be stored in a volume of approximately 10 m on each side. Third, like gasoline, uranium reserves are not infinite. For example, if the United States were to invest heavily in nuclear fission reactors, it would become dependent on other countries for its supply. There are, however, other possible methods of producing nuclear power. Plutonium 239 can be formed in fast breeder reactors from very abundant uranium 238, and this can be used in a fission chain reaction, similar in form to that of Equation (6.4), to produce energy. Here too there is a problem because the plutonium can be relatively easily converted to nuclear bombs. Finally, there are fears concerning nuclear reactor safety, although the deaths and injuries per year from energy production using hydrocarbon fuels are far greater than from nuclear fuels, even when weighted by the amount of energy produced by each process.

Some of the above problems would be alleviated if nuclear fusion reactions could be used to provide energy. Here very abundant deuterium, which is an isotope of hydrogen, could be used. Extremely high ignition temperatures are required to produce the reaction that converts the deuterium to helium. A breakthrough in fusion research would profoundly alter the world's energy outlook, but unfortunately, even optimistic proponents of nuclear energy do not see this occurring in the near future.

6.3 Renewable Energy

The ratio of the production time of hydrocarbon fuels (by means of photosynthesis and sedimentation) to that of the combustion time is of the order of 10^{13}; oil takes hundreds of millions of years to produce but only minutes to burn. This ratio for wood is around 10^6. For biomass energy such as the production of alcohol from sugar cane, the ratio of production to combustion time, which we will call R_{pc}, is around 10^5, and for animal dung it is one thousand or less. As R_{pc} becomes smaller and smaller, it becomes more and more feasible to renew the particular energy source. To consider oil as a renewable energy source implies a projection of millions of years into the future, and this is clearly absurd. Planting trees to produce firewood requires a projection of tens of years, and this is more realistic.

The notion of renewable energy implies that R_{pc} should be about 10^5 or less. Apart from the burning of trees and the production of fuel from biomass, other renewable energy sources are hydroelectric energy, wind energy, and direct solar energy. For the latter, which includes passive house heating, flat plate roof collectors, as well as thermoelectric, and photovoltaic devices, R_{pc} is close to or equal to 1. Here the energy (from the incident sunlight) is "produced" at the rate it is converted to applications.

For the highly industrialized countries the phrase "renewable energy sources" is often used interchangably with "alternative energy sources." However, in the less developed countries renewable energy is often the primary source of energy. For example, in Bangladesh 70 percent of the energy used is from fuel such as dung and rice residues. Even in the industrial countries the mix of energy sources varies. In Sweden nearly 60 percent of the electricity production is from hydroelectric power, a renewable energy source.

It is not possible in our introductory book that focuses on basic principles to do even a survey of the various methods of obtaining energy from renewable sources. For example, in order to explain photovoltaic conversion it would be necessary to outline the theory of semiconductors. Here I will present an analysis of one renewable energy source only, the windmill, because it is an excellent example of how basic fluid mechanics can be used to provide insight into performance and power output. I will then conclude this section by examining the available renewable energy supply.

6.3.1 An Example of a Renewable Energy Source: The Windmill

Windmills have been used to provide mechanical power for thousands of years, both in the East and in the West. They were most extensively used as prime movers in the seventeenth and eighteenth centuries, their use declining with the development of steam engines. Yet even in the ninteenth and the twentieth centuries they have played some role, particularly in the sparsely populated areas of Australia, Russia, and the United States, and in countries whose oil supply has been limited or cut off, such as occurred to Denmark during World War II. More recently there has been a renewed interest in large-scale power generation using windmills.

As in all first-cut engineering analysis, we will make rather drastic simplifying assumptions for the analysis of the windmill. First we will assume that the air flow is incompressible (a good assumption for air speeds less than 50 m/s (110 mph)) and inviscid (that is, frictionless). Second, the thrust (force) of the wind is assumed to be uniform across the windmill. Finally, we will assume that the windmill blades (often called the actuator) do not impart any rotation to the flow downstream. So downstream from the windmill the flow is assumed to be uniform and parallel. More refined engineering analysis lifts these assumptions, and as you may guess, shows that the efficiency we will derive here is an upper limit.

Figure 6.3 shows the side view of our rather abstract windmill. The air far upstream (where it is undisturbed by the presence of the windmill) has a velocity V_∞ and a pressure p_a. The objective of the windmill is to extract energy from the flow. The energy equation (Section 3.4) then suggests there will be a pressure drop across the windmill: On the diagram we have denoted the pressure p_1 and p_2, where $p_2 < p_1$ for the regions just upstream and downstream of the windmill respectively. The windmill will cause a wake; the velocity of the air downstream will be slowed down by its presence, just as any solid (or partially solid) body produces a wake. This will persist far downstream, even after the pressure has settled back to that of the surrounding atmosphere. The conservation of mass then requires that the flow spreads out behind the windmill because $V_\infty A_\infty = \overline{V}_3 A_3$ (we are assuming the air is incompressible). The

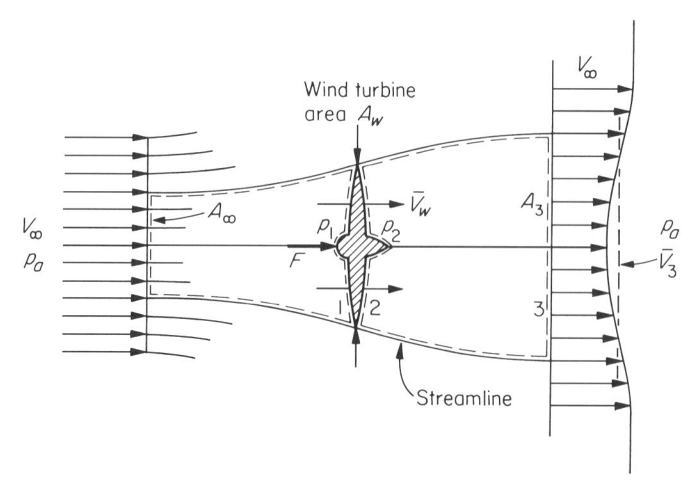

Figure 6.3 A sketch of a wind turbine. The mean wind speed is V_∞, and it is at a pressure p_a. The turbine slows the wind, thereby extracting energy from it. The dashed lines define the control surface.

areas A_∞ and A_3 are the left and right sides of the control volume respectively (Figure 6.3). The upper and lower surfaces of the control volume are defined by streamlines – imaginary lines that are tangential to the velocity vector so that no flow crosses them (see Section 3.3). Notice that I have written $\overline{V}_3$ with an overbar to denote an average value because (unlike V_∞) it varies across the wake.

We begin by applying the Bernoulli equation (Section 3.3) from V_∞ to the point just before the windmill. Note that since the flow is assumed to be inviscid and incompressible, there is no change in internal energy, so Equation (3.22) can be used. We then apply it again from just after the windmill to far downstream, where the pressure has returned to that of the atmosphere. (The Bernoulli equation cannot be applied across the windmill itself because there is a discontinuity of the streamline here.) These equations are

$$\frac{1}{2}\rho V_\infty^2 + p_a = \frac{1}{2}\rho \overline{V}_w^2 + p_1 \tag{6.5}$$

and

$$\frac{1}{2}\rho \overline{V}_w^2 + p_2 = \frac{1}{2}\rho \overline{V}_3^2 + p_a. \tag{6.6}$$

The velocity, $\overline{V}_w$, is the same just before and just after the windmill because mass must be conserved, and the area change is negligible. Eliminating p_a from these two equations yields

$$p_1 - p_2 = \frac{1}{2}\rho\left(V_\infty^2 - \overline{V}_3^2\right). \tag{6.7}$$

Notice that our windmill is similar to a gas turbine, with an inlet velocity (V_∞) and an exhaust velocity ($\overline{V}_3$). Both are like jet engines in reverse: They extract energy from the flow rather than add to it. In Chapter 3 we saw that the thrust force provided by a jet engine was (Equation 3.42)

$$Th = \dot{m}(V_{\text{out}} - V_{\text{in}}).$$

Here we rewrite this equation as

$$F = \dot{m}(V_\infty - \overline{V}_3) = \rho A_w \overline{V}_w (V_\infty - \overline{V}_3), \tag{6.8}$$

where F is the force imparted *to* the windmill. Note that I have written $\dot{m} = \rho A_w \overline{V}_w$. This is (by the conservation of mass) the same $\dot{m}$ as that entering at the position where we have defined V_∞ or that leaving at position 3 (Figure 6.3), because we are concerning ourselves only with the air between the streamlines. We have determined $\dot{m}$ at the windmill itself because here there is a well-defined area, that of the windmill, A_w. Equation (6.8) states that the force imparted by the air to the windmill is equal to the velocity deficit caused by the windmill times the mass flow rate. It really is an expression of the conservation of momentum because $\dot{m}V_\infty$ is the rate of momentum entering the control volume and $\dot{m}\overline{V}_3$ is the rate at which it leaves. Because they are not equal, Newton's second law implies that there must be a resultant force. The larger the difference, the larger the force.

Equations (6.7) and (6.8) can be combined to show that

$$\overline{V}_w = \frac{V_\infty + \overline{V}_3}{2}. \tag{6.9}$$

(To do this write Equation (6.7) as $p_1 - p_2 = \frac{1}{2}\rho(V_\infty + \overline{V}_3)(V_\infty - \overline{V}_3)$ and note that $F = (p_1 - p_2)A_w$.) Equation (6.9) shows that the air velocity at the windmill is the average of the free-stream velocity (V_∞) and the velocity in the wake, $\overline{V}_3$.

We now define the windmill interference factor, a, by

$$\overline{V}_w = V_\infty(1 - a). \tag{6.10}$$

If $a = 1$ the windmill stops all the air; if $a = 0$, there is no interference. Combining Equations (6.9) and (6.10) there results

$$a = 1 - \frac{V_\infty + \overline{V}_3}{2V_\infty}. \tag{6.11}$$

Notice that if all the wind energy is absorbed, then $a = \frac{1}{2}(\overline{V}_3 = 0)$.

The power imparted to the windmill is the mass flow rate times the change in kinetic energy across the control volume:

$$P_w = \rho A_w \overline{V}_w \left(\frac{V_\infty^2}{2} - \frac{\overline{V}_3^2}{2} \right). \tag{6.12}$$

(This can be formally seen to be correct by using the energy equation, Equation (3.33), with $\dot{Q} = \dot{W}_f = g\Delta z = \Delta p = \Delta u = 0$ across the control volume, and with $\dot{W}_s = -P_w$.) Using Equations (6.9) and (6.11), Equation (6.12) can be rewritten as

$$P_w = 2\rho A_w V_\infty^3 a(1-a)^2. \tag{6.13}$$

It is evident that the power is proportional to the cube of the air speed: A twofold increase in the wind will increase the output power eightfold. This is a most important result, as you will see by doing Problem 6.6. What value of a will provide the maximum power for a given wind speed and windmill area? Equation (6.13) shows that if a is either zero or unity, there will be no power output at all. In order to determine the maximum power, we differentiate Equation (6.13) with respect to a and set the result to zero:

$$\frac{dP_w}{da} = 2\rho A_w V_\infty^3 (1 + 3a^2 - 4a) = 0. \tag{6.14}$$

Thus, the maximum power occurs when $a = 1/3$. From Equation (6.13) its value, $P_{w\,\max}$, is

$$P_{w\,\max} = (16/27)((1/2)\rho A_w V_\infty^3). \tag{6.15}$$

This is known as the Betz limit. The total power of the wind crossing area A_w is

$$P_{\text{available}} = \frac{1}{2}\rho A_w V_\infty^3. \tag{6.16}$$

Thus, the maximum theoretical power developed by a windmill cannot be greater than $16/27 = 0.593$ (approximately 60 percent) of the power of the wind. It should be emphasized that this is for an ideal windmill, with no friction. It is quite a curious result because 16/27 is a rather complicated fraction to appear in an idealized analysis like this.

For $a = 1/3$, Equations (6.9) and (6.10) show that $\overline{V}_3 = 1/2\overline{V}_w$, that is, the average velocity in the wake is half that of the wind velocity at the windmill. This, from the conservation of mass, implies that the wake area is $2A_w$, or twice the area of the windmill. By placing the windmill in a shroud, the wake area can be increased. This causes the thrust of the activator to be increased. If, at the same time, $\overline{V}_w$ is maintained at $2/3V_\infty$, that is, the optimal value of $a = 1/3$ is still achieved, then the windmill can attain efficiencies somewhat higher than the Betz limit.

We have used the conservation of mass and the Bernoulli theorem to determine the maximum power that an ideal windmill can attain. When friction and transmission losses are included, the maximum available power will decrease. Equation (6.15) is a benchmark, a value to aim for, but of course a value that cannot be achieved. If you see a working windmill you can now at least calculate in a few moments the maximum theoretical power. You can then use this value to provide a rough estimate of its real output. Typically, most engineering systems work at efficiencies of 10 percent to 40 percent of their theoretical maximum, and a windmill is no exception. We will use such estimates in the next section.

6.3.2 The Availability of Wind, Sun, and Water

In the analysis of the windmill, we addressed the following question: Given an input (the wind), what is the maximum possible output shaft power available to drive an electric generator (or some other machine)? This is a problem of device efficiency. Similar analyses must be conducted for other energy conversion devices. For example, for a photovoltaic device the question to be addressed is: Given an input (solar electromagnetic energy in certain wavelength bands), what is the maximum amount of electrical power available? Here too an analysis of the device must be carried out. Although for the windmill knowledge of fluid mechanics is required, for the photovoltaic device the analysis tools are solid state physics and electromagnetic theory. As I have mentioned previously, we do not have the space here to develop these particular tools.

However, apart from determining the efficiency of energy conversion, there is yet another vital question that must be addressed: Given that it is possible to convert energy from one form to another (wind to turbine work, solar to electrical energy), is there enough of the energy source available to make it worthwhile? It is to this question that we now turn our attention.

At present the world population is approximately 5.6×10^9 and the per capita power consumption is about 2 kW. Thus, approximately 3.5×10^{20} J of energy is consumed each year. As we have emphasized, at present and for the forseeable future the bulk of our energy will come from fossil fuel. So the question we have just raised can be modified to: Can renewable energy significantly shift our energy dependence from fossil fuel sources? If calculations show that under optimum conditions and the highest conceivable efficiency, only 2 percent, say, of the world's total energy needs could be produced by solar or wind energy, then clearly its large-scale implementation will not cause much change. The CO_2 increase and other global pollution issues will not be alleviated because we will still have to rely on traditional fuel sources. If, on the other hand, solar or wind sources could potentially supply a large fraction of our power, then they should be carefully considered. Let us begin, then, by seeing whether there is enough power available from renewable energy sources to significantly shift the balance.

(a) The Wind. In Chapter 5 we showed that the total power of the atmospheric motion was approximately 2×10^{15} W. The present rate of the world's energy consumption is approximately 10^{13} W, two orders of magnitude less, or less than 1 percent of the total wind power. So in principle there is an abundance of wind energy. Of course this energy must be captured by means of windmills, and these can only be built, in practice, at the surface of the earth (rather than in the sky!) and on land. Preferably they should be located not too far away from where the energy is to be used. Locating them in central Canada or Antarctica is highly impractical. Moreover, even where siting is practical, they cannot cover too much of the land.

In assessing the practicality of the wind (or any other energy source) it is useful to start by doing a back-of-the-envelope calculation. If the result is encouraging, more refined calculations are then carried out.

Consider, for example, the energy needs of Australia. Its population is close to 18 million, and the per capita power consumption is about 6 kW (much higher than the world average of around 2 kW). Thus, the total energy consumption per year is about 3.4×10^{18} J, one hundredth of that of the world. Let us assume that on average the wind speed on the eastern seaboard is 7 m/s (around 15 mph). The question we ask is: What area of the land would have to be covered with windmills in order to supply 2.4 kW per capita or approximately 40 percent of the Australian energy needs? (This would cover the domestic needs and a good fraction of the industrial needs.) We must begin by making some intelligent guesses and assumptions. First, assume that we can make windmills with wind vanes of 30 m diameter. Although large, windmills of 50 m diameter have been successfully operated. Equation (6.15) then shows that the maximum power available (with $V_\infty = 7$ m/s) is 86 kW. If this windmill could be operated with, say, 20 percent efficiency, then it would supply approximately 17 kW of electricity. Forty percent of the total Australian energy needs is 1.4×10^{18} J per year. 17 kW amounts to 5.4×10^{11} J per year. So 2.6 million windmills would be needed. This is approximately one for every seven people. How much land would be needed? The span of the windmill is 30 m and we should probably allow 20 m between them, and a downwind distance of around 300 m, that is, approximately 50×300 or 15,000 m^2 of land is needed per windmill (Figure 6.4). So we could locate 67 windmills per square kilometer, and these would produce $67 \times 17 \times 10^3 = 1.14 \times 10^6$ W. In order to produce forty percent of Australia's power needs (approximately 2.4×10^3 W/person $\times 18 \times 10^6$ people $= 4.3 \times 10^{10}$ W), $4.3 \times 10^{10} \div 1.14 \times 10^6 = 3.8 \times 10^4\ (\text{km})^2$ of land would be required. This is a plot of 195×195 km (122×122 miles), less than 1 percent of the Australian land area. Of course the windmills would be distributed so there would be a number of small plots rather than one large one.

(a)

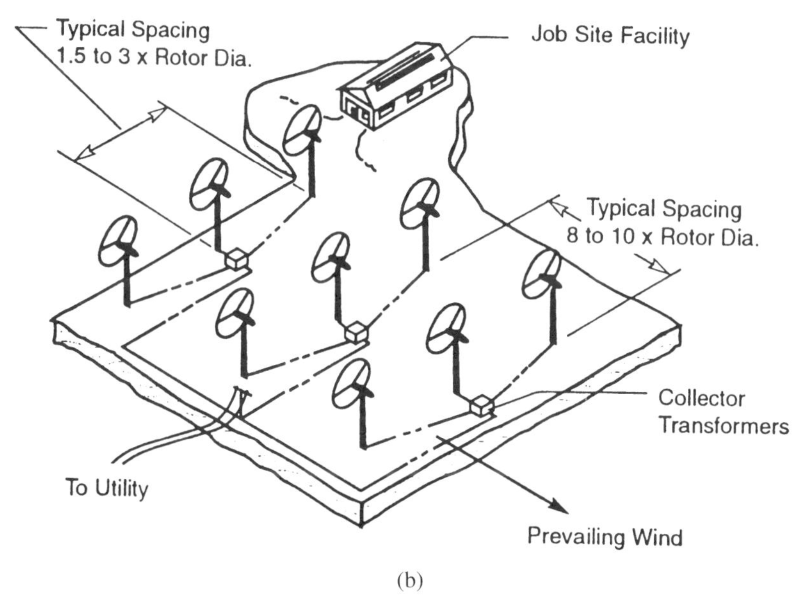

(b)

Figure 6.4 (a) A wind farm. (b) A sketch of an ideal wind farm showing the windmill spacings. (Part (a) courtesy of Paul Gipe. Part (b) reprinted from *Wind Turbine Technology, 1994* with the permission of ASME.)

Our calculation suggests that if the Australians became dedicated to windmill technology, much of their energy need could be supplied. Nearly three million windmills is a lot you may say. But there are more than that amount of vehicles, television sets, and so on. And moreover, a $195 \times 195\ \text{km}^2$ square plot, although large, is not impractical. There are missile test ranges that use more land than this.

Our crude calculation needs a further, more refined assessment. For example, can we rely on a 7m/s wind? Note that the power of a windmill goes as the cube of the wind velocity, so decreasing V to 3.5 m/s (a factor of 2) decreases the power by a factor of 8. Thus, the land area needed would grow by this factor: the plot would now have to be $550 \times 550\ \text{km}^2$. Further, even if the wind speed average were high enough, what about the effects of its intermittency? There is no such thing as a steady wind, constant at, say, 7 m/s, day in and day out. Sometimes it is much higher than the average, and sometimes there are long periods with no wind. Can the windmill tolerate variations in wind speed? Can the energy be stored to be used when there is no wind? Such questions are complex and in turn provoke further questions concerning material science (will the windmill blades stand up to rapid wind variations?) and socioeconomics (what are the energy usage patterns of society as a whole, and of various groups within it?).

In spite of these issues, our analysis suggests that wind energy is a possible alternative. If the average wind speed in Australia were only 1 mph (0.5 m/s) windmills would certainly be out of the question. For this speed the land area required would be prohibitive. Moreover, the friction losses would reduce the efficiency to close to zero. (Similarly, if the energy content of a hydrocarbon fuel were 1 MJ/kg rather than 50 MJ/kg, traditional automobiles would be out of the question because they would have to have 500-gallon fuel tanks to provide a reasonable range.) It appears, then, that for wind power, the overall supply is not the problem. It is the much deeper and more complex question that needs to be addressed: Under what conditions does society consider it worthwhile to convert its primary energy source from one form to another? We will address this question in Chapter 7.

(b) The Sun. In Chapter 5 we saw that the solar power at the mean earth distance from the sun is $1{,}370\ \text{W/m}^2$. Thus, the earth receives $1{,}370/4 = 343\ \text{W/m}^2$ distributed over its total surface area (Figure 5.2). Approximately one-third of this is reflected back into space by the clouds, atmosphere, and sea, so about $200\ \text{W/m}^2$ is available for collection. Let us return to our Australian example to determine if solar radiation is a viable alternative to hydrocarbon fuels.

We ask the question in the following way: If the conversion efficiency of solar energy to electricity is 5 percent, how much land area would be needed to supply 50 percent of Australia's annual energy needs? Here we have deliberately taken a low value for the solar conversion efficiency, which we define as the ratio of the electrical power output per square meter of land area to the power input from the sun per unit area (200 W/m^2). So we are assuming each square meter of land can produce 10 W. Hence to supply half the power needed per capita (3 kW), 300 m^2 of land would be required. And for the whole population of 18 million, an area of 5,400 km^2 (or a square of side 74 km) would be needed.

If we extend this to the U.S. population of 250 million people, the land area required to produce 4 kW per capita, or about 40 percent of our total power needs, would be around 100,000 km^2 (a square of side 316 km). As for wind power, our calculations are very crude. Nevertheless, the result suggests it is feasible to produce solar power in significant quantities. On the other hand, a population requiring similar power needs to ours, living on one of Jupiter's satellites, would be disappointed with their prospects of solar power, as you will show in Problem 6.8.

You may dampen the optimism that these calculations elicit by pointing out that countries such as England, where the sky is often gray, or Finland, where it is dark for nearly half the year, are poor prospects for the serious use of solar power. This, of course, is true. But the point is that there are many countries that *do* receive enough sunlight to provide a considerable amount of their total power needs. These include countries with very large populations such as the sub-Saharan Africa as well as parts of Central America, China, and the United States.

I will not be able to deal in any detail with the methods by which we convert the incoming solar energy into useful electrical power. Perhaps the most important recent advances in this subject are in photovoltaic technology. Here the incident sunlight produces a voltage in a diode (a semiconductor $p - n$ junction). The sunlight is focused on the device by means of mirrors or lenses. I was able to explain the principles of wind energy conversion (Section 6.3.1) because we had laid the fluid mechanical groundwork in Chapter 3. An equivalent amount of groundwork in electrical circuit theory and elementary quantum theory will enable you to understand the basic principles of photovoltaic technology, but these subjects are beyond the scope of our story of the engine and atmosphere.

Although the sun and wind show potential as energy sources for electrical power plants, they do not appear to be likely contenders as energy sources for vehicles. Although there has been some success with experimental cars covered with solar panels and with sails, it is difficult to imagine a production automobile with a windmill on its roof and a patient driver waiting for a wind gust so he or she can clear the traffic lights! Yet the energy of the wind or the sun can be converted to storable energy sources that can then be used to power automobiles. One of the most promising possibilities is the hydrogen fuel cell. Electrolysis is first used to separate hydrogen and oxygen from water. The hydrogen, with its immense chemical energy (Section 6.1), can then be used as a reactant, with oxygen, in an electrolytic cell. These cells can be used to power automobiles, which would carry on board hydrogen rather than gasoline. The principles are well understood, and fuel cell technology is advancing rapidly. The environmental and social issues of converting from one primary energy source to another will be discussed in the next chapter.

(c) The Water. The oceans, rivers, and lakes of the world have immense potential as energy sources. The wave action and tides of the seas can be used to drive generators, the temperature difference between the ocean surface and its lower depths can be used to operate heat engines (see Problem 6.10), and the potential energy stored in lakes and behind dams can be used to produce hydroelectricity.

Around 15 percent of the world's electricity is presently generated by hydropower. Here the potential energy of the water is stored behind a dam. It descends through a duct and converts its potential energy into shaft work. The turbine output is then used to power electric generators. In principle we have developed the equipment needed to analyze a hydropowered system. The energy equation for an open system (Equation (3.32) or (3.33)) provides the way of determining the shaft work (Problem 6.9), and the analysis of the water turbine itself is not too different from that of a windmill, although unlike most windmills, the water turbine is shrouded.

As for wind and sun, the use of water to provide power will undoubtedly increase. Although power from the tides and from waves is attractive, the technology of converting this to useful energy is difficult and is still in its infancy. Hydroelectric generation is well understood, and the efficiency of energy conversion is high. It is estimated that perhaps as much as three times the amount of hydroelectricity presently produced (which is around 20×10^{18} J per year) could be generated by hydroelectricity in the future. The present global total energy usage per year is approximately 350×10^{18} J. So hydroelectricity does already play a significant role. Ironically, it is environmental and social considerations that limit the larger-scale use of this "clean" source of energy. Thus, although the turbine and generator do not produce any emissions, the construction of large reservoirs can require mass population migration and cause damage to the ecosystem and unwanted changes in the local climate. There has been significant protest in recent years by the public and by environmental groups against the building of large hydroelectric systems, even in large countries such as Canada, Russia, and China.

6.4 Are Our Energy Sources Finite?

In the previous section we showed that in principle we have an abundance of wind, water, and solar power. If our planet were further from the sun, if there were no rivers or lakes, or if the winds were weaker, this would not be the case.

What about hydrocarbon fuels? Here too the world is abundantly endowed. There are over 200 years of proven coal reserves and over one thousand years of estimated coal resources. There are great abundances of shale oil and natural gas. All of these hydrocarbon fuels can be used in power plants or converted to fuels that are suitable for use in vehicles. Even for oil, the most sought-after fuel, there are proven reserves of over fifty years and geologists estimate that there is much more. In fact, the more prospecting that is done, the more hydrocarbon fuel that is found, be it coal, shale oil, natural gas, or oil. The situation is similar, too, for nuclear fuels. At present about 5 percent of the world's energy is derived from nuclear sources. If all these reactors used only uranium, and there were no recycling of the waste products, there would be enough uranium for over 300 years if used at the present rate. So even if 30 percent of the world's energy sources were from uranium, there would be enough for over fifty years. This assumes the most inefficient use of the uranium. By recycling the fuel efficiently, the uranium supply could last at least three times as long. And if fast breeder reactors are used, the supply is virtually limitless.

It appears, then, that the issue is not whether we have enough energy sources – we are over endowed with them – but what the cost is to the environment, and to society, of utilizing them. We will return to this issue in the next chapter.

6.5 Summary

In this chapter I have introduced you to the principal energy sources available to us, for our vehicles and for our power plants. We have looked at the subject from a broad perspective and can come to some general conclusions:

(i) There are three main categories of energy sources. These are (a) chemical reactions (or combustion), (b) nuclear reactions, and (c) direct utilization of wind, sun, and water. By far the most important of reactants in category (a) is hydrocarbon combustion in air, that is, the burning of oil, coal, gas, wood, and biomass. Approximately 90 percent of the present world energy use comes from these sources. Category (b) includes nuclear fission and nuclear fusion, although the latter is only at the experimental stage. Nuclear reactions are far more efficient than chemical reactions: Combustion produces a few eV per molecule of reactant in contrast to nuclear fission, which produces around one hundred million eV per atom (Section 6.2). Moreover, nuclear power production has no direct effect on the atmosphere, and in principle its waste can be stored without affecting the environment as a whole. Yet there are immense social problems associated with its use. In particular, there is fear of accidents and of weapons proliferation. Finally, (c), the wind, sun, and water may be directly used as a source of power. Notice that wind energy, solar energy, and hydroenergy are all derived from a nuclear source. We regard these forms of energy as clean because the nuclear power station, the sun, is a comfortable 150 million km away from us. It does not pose the same threat as a local nuclear reactor.

There are many hybrids of the above three categories. For example, hydroelectricity is often fed into the power grid to supplement coal-generated electricity, and solar heating often supplements gas heating in homes. These and other hybrids will become more important as we reduce fossil fuel combustion.

(ii) In principle there is no energy crisis. Even if our oil were to run out and we chose not to turn to nuclear power, there is enough coal, gas, wind, sun, and water to satisfy our needs, and we have the technology to do it.

Our energy problem is not due to lack of resources or technology; it is environmental and social. But here too, as you will see in the next chapter, the engineer has an important role to play.

6.6 Problems

6.1 (a) Sometimes detective work is required to determine the makeup of a hydrocarbon fuel from an examination of its exhaust products. A particular analysis of the products of combustion shows that they are 11.1 moles of CO_2, 9.4 moles of O_2, 0.9 moles of CO, and 99.45 moles of N_2. There is also some H_2O, but the amount was not measured. These products were formed by burning one mole of a hydrocarbon fuel, which is presumed to be of the form C_xH_y. Determine the formula of the hydrocarbon fuel and the number of moles of H_2O produced for 1 mole of C_xH_y burned.

(b) For the reaction in (a) determine the ratio of the mass of fuel to that of air (O_2 and N_2). This is known as the fuel-to-air ratio. Then determine the fuel-to-air ratio if the reaction were stoichiometric; that is, if the same molecule produced only CO_2 and H_2O (and of course N_2) as its products. Finally, determine the ratio of the actual to stoichiometric fuel–air ratios. This is called the equivalence ratio, Φ.

6.2 (a) Assuming that it requires as much energy to create an elementary CH_2O molecule (Equation (6.2)) as is given up when the biomass is burned (Equation (6.1)), estimate how many photons are required to produce one CH_2O molecule. The energy of a photon of energy is hf, where h is Planck's constant and f is the frequency of the radiation (see Section 4.3). For photosynthesis to occur, the wavelength of light must be in the 1-μm range.

(b) A coniferous forest stores about 50×10^6 J/m^2 of solar energy per year by converting sunlight into biomass. Show that this is less than 1 percent of the average incident solar power. How many kilograms of coniferous wood could be grown per year in a 5×5 km forest? How many people could this supply at a rate of 3-kW per person per year? How big would the forest have to be to supply the whole U.S. population at this rate? What percentage of the total U.S. land area is this?

6.3 The atmosphere contains a small trace of hydrogen that has not yet escaped to space. Its concentration is approximately 0.5 parts per million by volume of the atmosphere (ppmv). What is the total mass of hydrogen in the atmosphere? If the U.S. consumption of energy is 80×10^{18} J/yr, how long would this source of energy last if it could in some way be extracted from the atmosphere?
Do the same calculation for methane. Its atmospheric concentration is given in Table 5.1. The energy content of hydrogen and methane is given in Section 6.1.

6.4 The products of nuclear reactions are generally radioactive. Many of these products lose their radioactivity quickly, but others remain active for very long periods, giving off harmful ionizing radiation. For example, cesium 137 and strontium 90 have half-lives of 30 and 28.8 years respectively. Here the numbers 137 and 90 refer to the total number of neutrons and protons.
The expression "half-life" refers to the time it takes for half of the nuclear material to decay. The rate of decay, dN/dt, is proportional to the number of atoms, N, present at a particular time, t, that is,

$$\frac{dN}{dt} = \lambda N.$$

Here λ is the constant of proportionality. Solve this equation to show that the decay of N is exponential. Denote the number of atoms at time $t = 0$ as $N = N_0$. Show that the half-life, $T_{1/2}$, is $-0.693/\lambda$.
Consider 1 g of cesium 137. How many atoms are there? (Avogadro's number is 6.02×10^{26} atoms/(kmol).) Given that the half-life of cesium is 30 years, determine how many disintegrations there are per second. How long will it take for 99 percent of the cesium to decay?

6.5 In our analysis of the windmill (Section 6.3.1) we did not determine its overall thermodynamic efficiency. Show that the windmill is in fact only one component of a thermodynamic heat engine cycle. What are the high and low temperatures for this cycle? Sketch the thermodynamic system. Do you expect that it has a high thermal efficiency? Does it matter?

6.6 Consider a windmill with rotor diameter of 40 m and an efficiency of 30 percent. The average wind speed on two consecutive days of its operation is the same, 7 m/s. However, on day one it is steady at 7 m/s all day, but on day two it is 2 m/s for 8 hr, 12 m/s for the next 8 hr, and 7 m/s for the remainder of the day. On which day is the output the greatest?

Clearly the distribution of wind velocity is as important as the mean speed itself. Does your answer imply that a variation in wind speed about a mean value is better than a constant wind speed of the same mean value (i.e., gustiness is to be preferred to constant wind conditions)? Are there factors that might mitigate your conclusion?

6.7 The U.S. Department of Energy has developed a solar power plant in the Mojave Desert in California. It consists of 2,000 motorized mirrors that focus the sun's rays on a 91-m metal tower containing molten nitrate salt. The salt heats to a temperature of 565°C and then heats water to drive a 10-MW steam turbine that drives electricity generators.
   (a) Assuming a yield of only 10 W/m^2, how large would the plot of land need to be to provide the 10 MW? If the average daily electricity need of a Californian is 3 kW, how many people would be supplied?
   (b) What is the maximum (Carnot) efficiency of the power plant assuming the ambient temperature is 32°C. The actual efficiency of the plant will be much lower. Explain why.
   (c) The water flows through the turbine as steam. If the diameter of the turbine is 1 m, determine the approximate mass flow rate of the steam flowing through it to produce the 10 MW. For your order-of-magnitude calculation you may assume the specific heat, c_p, of steam is constant with a value of 2×10^3 J/(kg K). The inlet and outlet temperatures to the turbine are 450°C and 330°C respectively. Assume the contributions from the change in kinetic energy and potential energy are negligible and that the water vapor is a perfect gas (use Equation 3.33).

6.8 The Jovian moon, Io, is 5.2 times the distance from the sun as our earth is. What is the incident solar flux (W/m^2)? Assuming a 10 percent conversion efficiency, how large would a solar collector have to be to supply 10 kW of power to support a band of adventurous astronauts? What is the land area needed for this amount of power on earth?

6.9 Consider a hydroelectric power system. The height of the water in the dam is 100 m, and the flow rate through the turbine is 3×10^4 kg/s (4.75×10^5 gallons/min). The diameter of the duct is 2 m, and density of the water is 1,000 kg/m^3. If the water starts from rest (at the top of the dam) and it exits from the turbine to atmospheric pressure, use the first law of thermodynamics for an open system to determine the power output of the turbine. Assume that the flow is adiabatic and that 10 percent of the shaft work is lost to friction.

6.10 In order to construct a heat engine, a temperature difference is required. The second law of thermodynamics shows that the efficiency of the engine must decrease as the temperature difference decreases. Nevertheless, if there is a naturally occuring temperature difference, then it may be worth exploiting, no matter how small it is.

The upper levels of the tropical ocean are around 300 K, while at lower depths, where the water circulates from the arctic, the temperature is around 275 K.

(a) Assuming a heat engine could be constructed to exploit this temperature difference, what would be its maximum possible efficiency?

(b) The heat engine cycle could be accomplished in the following way. The working fluid would be heated near the ocean surface by circulating it past the warm sea water (in a heat exchanger). It would then pass through a turbine that would produce work and drive an electric generator. The temperature of the working fluid would drop as it passed through the turbine. It would be further cooled by the cold sea water at the lower depths and then pumped to the surface again to complete the cycle. The working fluid remains isolated from the sea water. A sketch is shown in Figure 6.5. The closed cycle is the same, in principle, as the turbine power generation system shown in Section 3.5 (Figure 3.16). However, for the gas turbine heat is provided by burning fuel, whereas for the ocean engine it is provided by the naturally occuring temperature gradient.

Consider a 100-MW power plant. Assume that its thermodynamic efficiency is one third of the Carnot efficiency determined in (a). The turbine delivers 7 kJ/kg of working fluid, and the pump requires 0.4 kJ/kg of working fluid.

What must be the mass flow rate of the working fluid? How much heat per unit time must be added to the fluid at the heat exchanger? Use the convective heat transfer equation, Equation (4.16), to provide an estimate

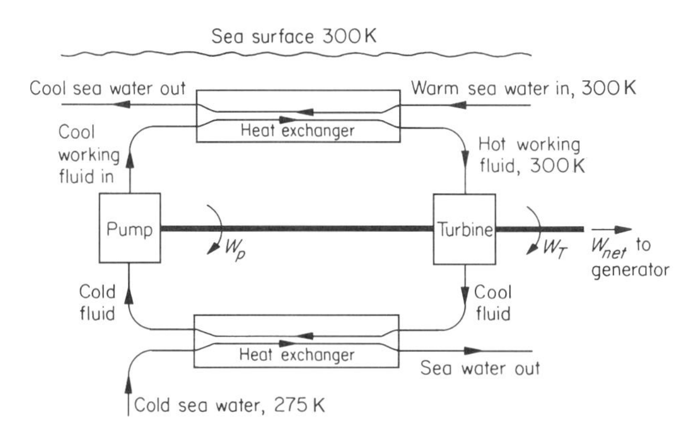

Figure 6.5 (Problem 6.10.)

of the area of the heat exchanger needed. Assume ΔT is 8 K and H_c is 500 W/(m^2 K). How many football fields is this? Why is it so large?

6.11 (a) One of the conclusions of this chapter is that we do not have an energy crisis: world-wide there is an abundance of fossil, nuclear, and renewable energy sources. Use your library and other sources to independently check this. Find out the world resources of coal, oil, natural gas (etc.) and determine how long these would last in terms of present day energy usage. Also determine the potential there is from renewable energy sources. Do you agree with my conclusion?

(b) Now consider the reserves of fossil and nuclear energy and the potential of renewable energy sources in the following countries: China, Japan, Italy, and the United States. What would the energy outlook for these countries be if they were to become completely isolated? Also find out the increase (or decrease) of energy consumption (both total and per capita) in each of these countries over the past 20 years, and comment on the trends.

Symbols

A_w	area of windmill	m^2
a	interference factor	
F	force	N
$\dot{m}$	mass flow rate	kg/s
P_w	power imparted to windmill	W
p_a	atmospheric pressure	Pa
p_1	pressure just before windmill	Pa
p_2	pressure just after windmill	Pa
p	pressure	Pa
Th	thrust	N
V	velocity	m/s
$\overline{V}_3$	average velocity across windmill wake	m/s
$\overline{V}_w$	wind velocity at windmill	m/s
V_∞	undisturbed wind velocity upstream of the windmill	m/s
ρ	density	kg/m^3

7

The Engine, the Atmosphere, and the Engineer

Throughout this book I have stressed concepts that are needed to understand both the engine and the atmosphere. These include the principles of thermodynamics, fluid dynamics, and heat transfer. I have also shown that the atmosphere and the engine are intimately linked: The exhaust from our engines changes the global and local characteristics of the atmosphere. So the subject of combustion, and energy sources in general, also plays a major role in the study of the engine and atmosphere. In this final chapter I will return to the overview presented in Chapter 1. But first I will review what we have already done.

7.1 Review

The second law of thermodynamics (Section 2.10) states that a heat engine must operate between a high-temperature heat source and a low-temperature heat sink, that is, there must be a temperature difference. The low temperature is usually the environment. The high temperature is achieved by burning some type of fuel or by direct energy from the sun. No heat engine can achieve 100 percent efficiency; if it could there would be no need for the low-temperature sink and the high-temperature source could be the environment itself (Section 2.10). We would not have to burn fuel to run our engines. There would be neither an energy nor an environmental "crisis."

There are devices that produce work that may appear to be free from the constraints of the second law of thermodynamics. In our analysis of the windmill (Section 6.3.1) we were not concerned with its thermodynamic efficiency. That is because we considered only one part of the heat engine cycle, the wind turbine itself. But the winds that power it must be generated, and here the energy source is the sun, which creates the temperature difference that drives the wind. In Chapter 5 (Section 5.1) I showed that the atmosphere is particularly inefficient in turning the sun's energy into wind. But of course this does not matter. The winds do not cost us anything, nor do they pollute. Similarly, the sun must evaporate the water that forms the clouds that provide the rain to fill the dams that drive hydroelectric plants. Here too the overall thermodynamic efficiency is very low, but it does not matter.

Because we need a temperature difference to run our engines, we need fuels. These invariably affect our environment in some way. Therefore, we need to use them sparingly. This implies we must try to design engines more efficiently.

The stuff inside our engines is usually a fluid (for example, air and gasoline vapor in automobiles, steam and water in coal-powered plants, and liquid sodium in nuclear reactors); therefore, we must understand its motion. In Chapter 3 we showed that friction plays a major role in fluid motion. We also showed that at high Reynolds numbers fluid flow becomes turbulent, and here the effects of friction become even more pronounced. Friction and turbulence reduce engine efficiency. However, turbulence also enhances mixing rates. So it has two sides, one beneficial and the other detrimental. Both are a result of the random eddy motion that occurs at a multiplicity of scales. These turbulent eddies very efficiently transport momentum, causing drag in pipes and reduced engine efficiency. They also transport quantities such as heat and mass, causing efficient mixing of reactants in engines and dispersion of pollution in the atmosphere. The actual mechanics of the turbulent mixing, be it of momentum, temperature, moisture, or mass, are similar.

In engines fluids are involved in the production of the high temperatures as well as in the heat transfer processes by which energy is transported from one part of the engine to another. Similarly, in the atmosphere heat is transferred from the equator toward the poles, as well as in the vertical direction, by the air motion. In Chapter 4 we studied the various mechanisms of heat transfer: conduction, convection, and radiation. There we stressed that when a fluid is in turbulent motion, the heat transfer rate is generally magnitudes greater than for laminar flow. We also saw how the atmosphere changes the rate of heat transfer from the sun to the earth, and the rate of heat transfer from the earth to space, by means of its radiation absorption.

In Chapter 6 we examined the various sources of energy. We described three broad categories: chemical reactions (combustion), nuclear reactions, and direct conversion (wind, solar, and water power). For chemical and nuclear reactions there are products of reaction. Ideally, the products of nuclear reactions can be confined, but those of combustion cannot: The 8 kg (18 lbs) of CO_2 produced by every gallon of gasoline burned ends up in the atmosphere (Section 1.4.1). There are also other gases (and particulates) produced by fossil fuel combustion, and a small fraction of these too end up in the atmosphere. In Chapter 6 we also showed that the world is abundantly endowed with energy sources. It is the environmental and social implications that present the major problems.

It is the fact that the atmosphere is affected by the products of combustion that motivated our study of it in Chapter 5. There we saw that the absorption of infrared radiation by the atmosphere is very strongly dependent on the amount of water vapor, carbon dioxide, methane, and other gases. By increasing the amount of these gases in the atmosphere, we may be affecting the global climate. The greenhouse effect (Section 5.2) plays a central role in the discussion of the engine and atmosphere. But the region very close to the ground, the atmospheric boundary layer, can, under inversion conditions (Section 5.3), trap other constituents such as nitrogen oxide and carbon monoxide. These gases (like CO_2) are produced in the exhaust of our cars and power plants. Although their concentration is very low, they cause smog and foul air. So they cause local rather than global problems. Although we have studied the meteorology of temperature inversions, we have not discussed how smog is produced. We will do so in Section 7.2.

I have particularly emphasized the integrated nature of the thermal fluid sciences. If I were writing a larger book that included solid mechanics and materials engineering, I would have shown how these are also integrated with other parts of engineering. The modern engineer must pull together many subdisciplines in order to design a new engine or power plant, or to analyze an environmental problem. Because of the increasing interrelatedness of technology, society, and the environment, in the future there will be less and less emphasis on high specialization, and more on integrated approaches. This does not mean that our knowledge will be shallower. It means that we will have to discriminate more carefully between detailed empirical knowledge and knowledge that is based on broad principles that, when properly mastered, can be applied with imagination to broad classes of problems.

7.2 Local and Global Issues: Small and Large Numbers Revisited

In Chapter 1 (Section 1.2) the significance of multiplying very small numbers by extremely large ones was emphasized. I showed that although each automobile exhausts a negligible amount of CO_2 compared with the atmospheric mass, when we add up the total contribution from all the internal combustion engines and power plants in the world, the increase of CO_2 concentration is measurable (Figure 1.6). In Section 5.2, on the greenhouse effect, I suggested that this increase could be causing global climate change. In this section I will stress again the effect of multiplying very small numbers by very large ones by examining local issues: the role of by-products of combustion on the environment. I will then briefly return to the subject of global climate change.

(a) Local Issues. We begin by recalling the equation for the combustion of octane, which we use as a representation of gasoline fuel (Equation (1.2)):

$$C_8H_{18} + 12.5(O_2 + 3.76N_2) \rightarrow 8CO_2 + 9H_2O + 47N_2. \qquad (7.1)$$

Precisely 12.5 moles of air (O_2 + $3.76N_2$) must combine with one mole of C_8H_{18} to produce CO_2 and H_2O as the only products of combustion. (The N_2 plays no role in this ideal reaction.) However, the combustion process is never perfect: Some of the gaseous fuel mixture hides out in oil layers, in cool regions near the cylinder walls, and in micro-crevices in the piston or cylinder. On occasions the engine misfires. In these cases some of the hydrocarbon fuel escapes combustion, and so it appears in the combustion products. These include methane and ethylene as well as many other molecules, of the form H_nC_m, that have lower molecular weight than the original octane. They are usually lumped together as a group and called unburned hydrocarbons (HCs). Moreover, as we have discussed in Chapter 6, when there is an excess of fuel over air, then carbon monoxide (CO) will also be produced.

Equation (7.1), then, should be written in the form,

$$C_8H_{18} + x(O_2 + 3.76N_2) \rightarrow aCO_2 + bH_2O + 3.76xN_2 + cCO + dHC. \qquad (7.2)$$

Here x will generally be different from 12.5. (The combustion is called rich if $x < 12.5$, lean if $x > 12.5$, and stoichiometric if $x = 12.5$.) We should hope that c and d are small compared with a and b, that is, that CO_2 and water are the dominant products of combustion, and this is generally the case. Typically, a U.S. car produces around 10 g of CO per km traveled compared with around 300 g/km of CO_2. Although CO_2 is a greenhouse gas, unlike CO and HCs it does not produce immediate health and environmental damage.

Yet even if we could achieve perfect stoichiometric conditions, there is still another problem. Although the nitrogen in the air is relatively inert at low temperatures, it undergoes chemical reaction at high temperatures. And of course our engines must run at high temperatures: The second law of thermodynamics shows that the higher the temperature the heat source is, the higher is the engine efficiency. So far I have neglected the role of atmospheric nitrogen, assuming it remains unchanged throughout the reaction.

At high temperatures the N_2 and O_2 in the engine dissociate and two chain reactions result:

$$O + N_2 \longleftrightarrow NO + N \qquad (7.3)$$

$$N + O_2 \longleftrightarrow NO + O. \qquad (7.4)$$

These equations are known as the Zeldovich mechanisms of NO formation. They may go in either direction, but an analysis of the reaction rate mechanisms shows that net NO results. There are also other chemical reactions that more directly involve the hydrocarbon fuel and further enhance the NO production.

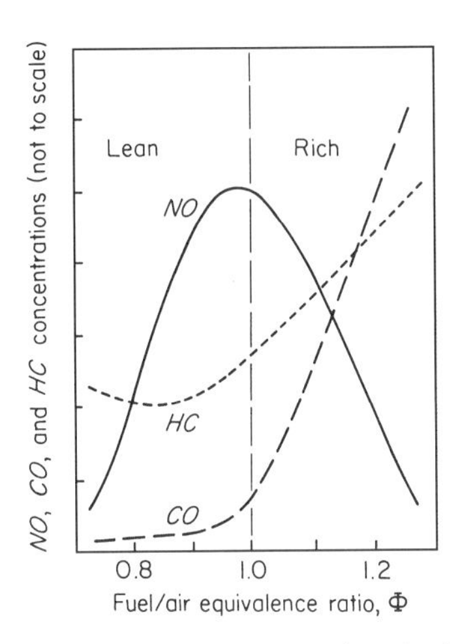

Figure 7.1 The HC, CO, and NO concentrations in the exhaust of a spark ignition IC engine as a function of the equivalence ratio Φ. Φ is defined as the actual fuel-to-air ratio divided by the stoichiometric fuel-to-air ratio. If it is greater than 1, the mixture is rich (excess fuel); if it is less than 1, it is lean (excess air).

For example, in the combustion process OH molecules (called hydroxyl radicals) are formed as a result of the dissociation of the air and fuel. These also combine with nitrogen to produce NO.

Figure 7.1 shows the way the exhaust gas concentrations of HCs, CO, and NO, in a conventional spark ignition IC engine, vary with the fuel-to-air equivalence ratio, Φ. This is ratio of fuel to air (F/A) in the actual mixture divided by the fuel-to-air ratio for stoichiometric conditions, that is, $\Phi = (F/A)_{\text{actual}}/(F/A)_{\text{stoichiometric}}$. (In Problem 6.1, I asked you to determine Φ for a particular reaction.) Rich mixtures ($\Phi > 1$) produce enhanced CO and HCs, but even for lean mixtures there are HCs and some CO. Notice that the peak NO formation is for approximately stoichiometric conditions. Thus, under all operating conditions there will be some CO, NO, and HCs. Coal power plants also produce these constituents (as well as SO_2 and SO_3, which cause acid rain, because of small amounts of sulphur in the coal).

The effects of minor exhaust products can be catastrophic. Carbon monoxide is a highly toxic gas. Levels of 0.1 percent concentration by volume in air cause death, and even 0.01 percent causes headaches and loss of mental acuity. Yet once the exhaust enters the atmosphere there are further reactions that occur in the presence of sunlight. In one sequence of reactions ozone (O_3) is produced. It can cause eye irritation, respiratory illness, and reduced pulmonary function. It can also reduce crop productivity (killing some types of plants and trees) and degrade materials. It is a principal constituent of what is called photochemical smog.

Ozone is formed by a series of photochemical reactions, that is, reactions that require radiation for them to proceed. Photosynthesis, the formation of biomass from water and carbon dioxide (Equation 6.2), is a photochemical reaction that you are familiar with. It cannot take place in the absence of sunlight. The rate of a photochemical reaction is determined in part by the frequency of the radiation. If the frequency band of the radiation is not right, the reaction will not take place at all. For example, photosynthesis does not occur in the dark, even in a heated room where there is infrared radiation. It requires higher-energy radiation, in the visible frequency range.

The way ozone is formed in the lower atmosphere is as follows. The NO from the engine undergoes a photochemical reaction with the unburned hydrocarbons (such as ethylene and butane) to produce NO_2.

$$HC + NO + hf \rightarrow a NO_2 + \text{other products} \tag{7.5}$$

Here hf is a quantum of solar radiation (h is the Planck constant, Section 4.3.1) at a frequency, f, greater than approximately 7.3×10^{14} Hz, that is, a wavelength less than approximately 0.41 μm. You may see, by returning to Figure 4.16, that there is a window that allows these wavelengths to reach the earth's surface. On the other hand, wavelengths less than about 0.3 μm are absorbed by the atmosphere. We will return to this point in Section 7.2(b). There is then a further photochemical reaction which forms NO plus O:

$$NO_2 + hf \rightarrow NO + O. \tag{7.6}$$

Here, too, the radiation wavelength required is less than 0.41 μm. The O then combines with O_2 to form O_3.

$$O + O_2 + M \rightarrow O_3 + M \tag{7.7}$$

Here M represents any molecule that removes excess energy from the reaction without taking any part in it.

In fact the O_3 formation is more complex than this. There are reactions involving the hydroxyl radical (OH) and other constituents that enhance the O_3 concentrations. There are also reverse reactions in which NO_2 and O_2 are formed from NO and O_3. The main point is that O_3 is formed close to the ground by a photochemical process requiring a combination of sunlight, HCs, and NO_x (NO and NO_2). That is why its concentration is high in regions of high automobile traffic density (although low concentrations occur naturally). The highest concentrations of O_3 occur when the exhaust can be trapped close to the ground, and this as we have seen in Section 5.3.1 occurs when there are inversion conditions. The sunlight turns the atmosphere above Los Angeles or Mexico City into a huge chemical reactor.

Let us now return to the issue of small and large numbers. Figure 7.2 shows a graph of HCs, NO_x, and CO emissions. These estimates were determined by the U.S. Environmental Protection Agency (EPA) for U.S. vehicles. Notice the

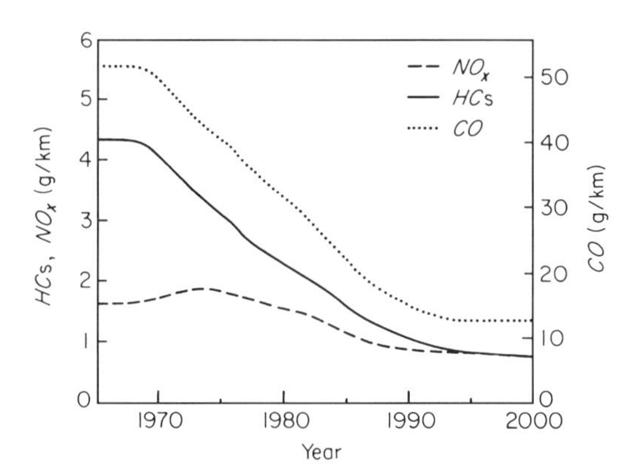

Figure 7.2 The HC, CO, and NO_x emissions for the average U.S. car since 1965. (U.S. Environmental Protection Agency.)

units used: g/(km driven). The reduction since 1965 (when emission control was implemented) has been dramatic. CO emissions have been reduced from over 50 g/km to around 12 g/km, and hydrocarbon emissions have been reduced from over 4 g/km to less than 1 g/km. These are for the average car on the road. New cars have emissions much lower than this: 2 g/km of CO, 0.25 g/km of HCs, and 0.5 g/km of NO_x can be achieved.

Figure 7.2 also shows that although the decrease in emissions has been very significant, there is a leveling off. The design of emission-limiting devices (such as catalytic converters), like many other areas of technology, develops quite rapidly at first, but then gains become difficult.

Unfortunately, during the period 1965 to 1990, urban miles traveled went up by 100 percent, so that gains in emission control have been largely offset. As a consequence air quality in the major cities has not changed very much. Figure 7.3 shows the U.S. emission of NO_x. Although the decrease in emission *per km traveled* was significant in the 1970s and 1980s (Figure 7.2), the *total* decrease was very modest because of the increase in total kilometers traveled (Figure 7.3). The issue of small numbers multiplied by large ones looms large in the problem of emission control.

When we examine global trends the problem looks even worse. Figure 7.4 shows the global motor vehicle registrations since 1930. The number of vehicles is growing much faster than the world population: between 1960 and 1990 there was 5.2 percent growth per year for vehicles and around 1.6 percent for population. Most of the vehicle growth is occurring in Asia and Latin America. Many of these countries have only mild, if any, emission controls. For example, in 1990 the CO emission per km traveled was 30 g for Western Europe and over 60 g in Eastern Europe and the less developed countries. It was around 12 g in the United States. So in spite of advances in emission control technology, the amount of pollution is increasing, particularly in the developing countries.

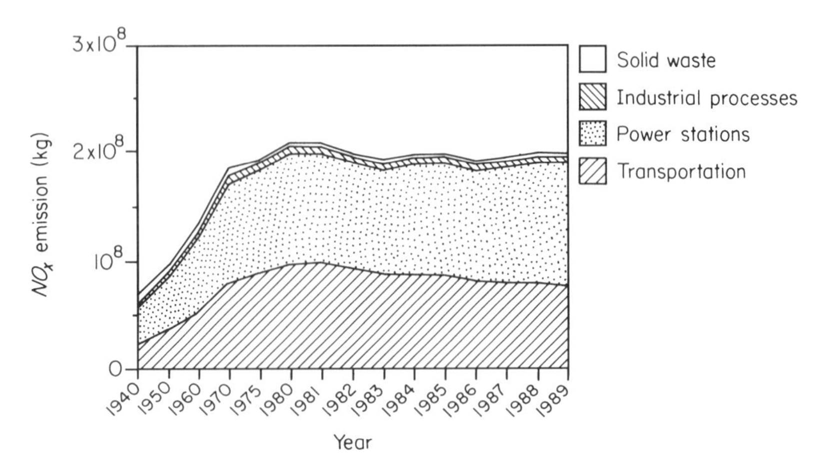

Figure 7.3 The U.S. NO_x emissions since 1940. (U.S. Department of Energy.)

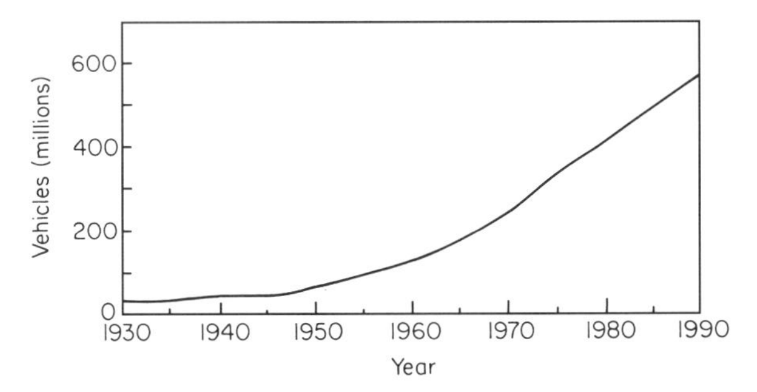

Figure 7.4 The number of vehicle registrations in the world since 1930. This includes cars, trucks, and buses (but not motorcycles). (After Green and Santini, 1995.)

It is difficult to determine accurately by how much the global budgets of CO, NO_x, and HCs have increased because emissions vary from car to car and their concentrations are determined by local atmospheric conditions. There are strong indications that their rate of increase approximately follows that of the population. For example, methane (CH_4), which is a by-product of both fuel burning and farming activity, seems to be following the population curve (Figure 7.5). It appears, then, that if vehicle production increases at roughly its present rate, and that if automobiles continue to use hydrocarbon fuels, then even if vehicle emissions were globally reduced by a factor of 10 from the predicted U.S. emissions of the year 2000 (Figure 7.2), a very unlikely possibility, worldwide pollution of our cities will still increase, because of the overwhelming increase in total vehicle miles traveled.

In Bangkok, Thailand, a rapidly developing Southeast Asian country, the city traffic system is so congested that the average speed of car travel is less than 10 km/hr (6 mph) day or night. Motorists spend approximately 44 days per year

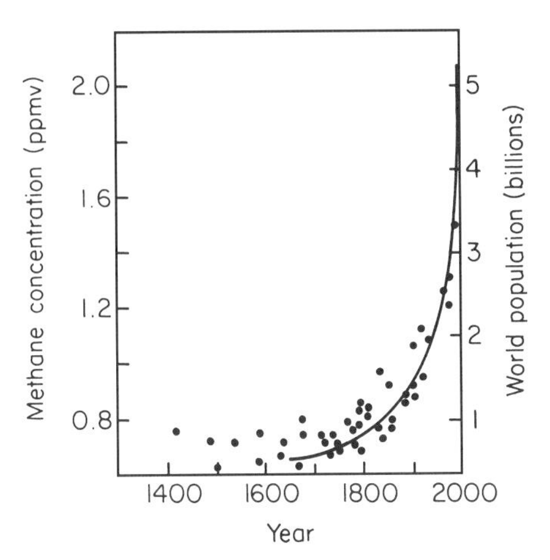

Figure 7.5 The world population (solid line) and the global methane concentration (dots) since 1400. (After "Trends 93," CO_2 Information Analysis Center.)

just idling in standstill traffic, costing over one million dollars in wasted fuel per day. At least a million people in Bangkok (population, 10 million) received treatment for smog-related respiratory problems in 1990. Similar stories can be told for Mexico City, Beijing, or Santiago.

(b) Global Issues. The minor fossil fuel emissions that we have just discussed have a direct, detrimental effect on our health. These constituents are not fundamental to the combustion process: In principle if we burned pure O_2 with pure hydrocarbon fuel we could avoid NO_x, CO, and HCs emissions. In practice, as we have indicated, even the smallest amounts of these constituents in the exhaust can have profound effects. But CO_2 must always be present in fossil fuel combustion. As we have seen in Chapter 1, its atmospheric concentration has increased from around 315 ppmv in 1955 to over 350 ppmv today (Figure 1.6).

Carbon dioxide is not the only greenhouse gas that results from our industrialization. The atmospheric absorption graph (Figure 4.16(b)) shows that methane (CH_4), nitrous oxide (N_2O), ozone (O_3), and CFCs, or chlorofluorocarbons, absorb long-wave radiation. In fact, per molecule, chlorofluorocarbon is ten thousand times more effective in absorbing long-wave radiation than CO_2. Chlorofluorocarbons have been banned primarily because of their effect on the stratosphere, yet the levels of other greenhouse gases are rising as the world population increases.

Unlike the local pollution problem, each country affects every other country with its greenhouse gas emissions. Although severe smog mainly affects residents of the cities that create it, the atmospheric loading of greenhouse gases affects us all. Cleaning our local environment does not get rid of the problem.

The dominant role of the wealthy countries such as the United States, Japan, and the countries of Western Europe in greenhouse gas production is evident

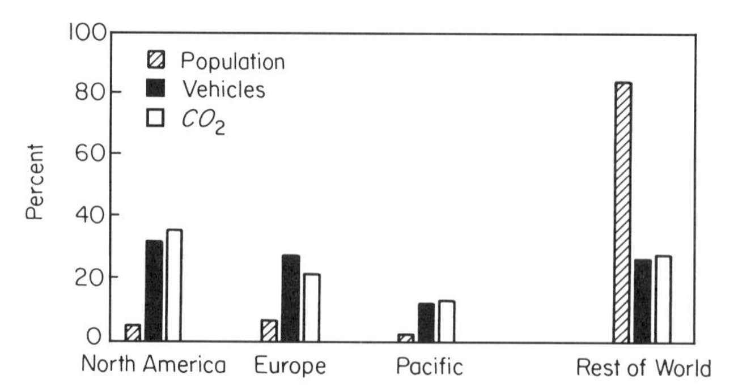

Figure 7.6 The present distribution of population, vehicles, and CO_2 production. The Pacific countries include Japan, Australia, and New Zealand. The rest of the world includes most of Asia, Africa, and Latin America. (After Greene and Santini, 1995.)

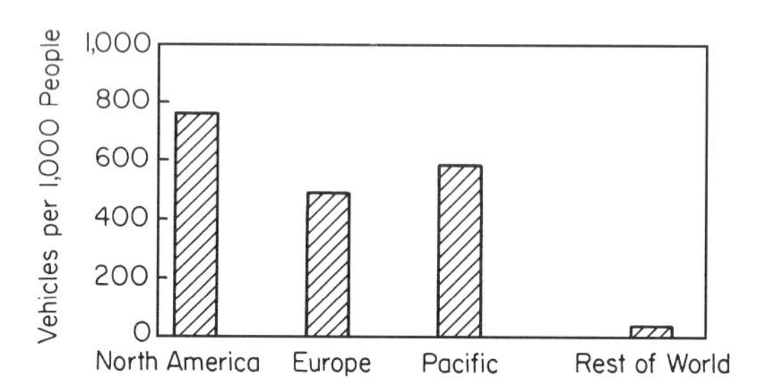

Figure 7.7 The present vehicle distribution per 1,000 people. In the United States there is nearly one per person. In Asia, Africa and Latin America it is less than one vehicle for twenty people. (After Greene and Santini, 1995.)

from Figure 7.6, which shows the global shares of population, vehicles, and carbon dioxide. The rich countries, constituting less than 20 percent of the world population, contribute over 70 percent of the global CO_2. This graph might suggest that these richer countries, who created the problem, are in the position to clean it up with their technology, which is the most advanced. They could institute alternative transport systems, more efficient power-generating plants, and so on. Yet when we look at the global distribution of vehicles per capita (Figure 7.7), we see that there is an even greater problem, and it is not confined to the wealthy countries.

Although in the United States and Europe there is approximately one vehicle for every one or two people, the rest of the world (which is four-fifths of the global population) has less than one vehicle for 20 people (Figure 7.7). This is rapidly changing. The growth rate of vehicle ownership is increasing fastest in the developing countries such as Thailand, Indonesia, Brazil, and Mexico. In some of these countries the number of vehicles is expected to triple in the period

1990 to 2010. All of these cars will add to the global CO_2 budget. The effects of increased fuel efficiency and alternative transport systems on greenhouse gases instituted by the richer countries will be insignificant compared to the growth in greenhouse gases from the developing countries.

There is yet another global issue that we have so far not addressed. It is the destruction of the ozone layer in the stratosphere. You will recall (Section 5.1.4) that the stratosphere is the region just above the troposphere. Here there is a temperature inversion, that is, the temperature increases with height (Figure 5.12). Because at these heights (approximately 10 km and above) the air is very thin, there is little absorption of solar radiation, and so very short wavelengths ($<0.3\,\mu$m) can penetrate into the stratosphere. Under these conditions oxygen dissociates by means of the following photochemical reaction:

$$hf + O_2 \rightarrow O + O. \tag{7.8}$$

Then the reaction shown in Equation (7.7) takes place to form O_3. Thus, the formation of ozone in the stratosphere is relatively straightforward. It does not require the presence of trace gases. On the other hand, we saw that in the troposphere, near the ground, ozone formation is a complex process indeed, requiring HCs, NO, and OH radicals. The principal reason for the difference is that the stratospheric photochemical reaction (Equation (7.8)) cannot occur near the ground because the very short wavelengths required for it to take place are absorbed by the atmosphere above (Figure 4.16). The inversion in the stratosphere is in fact caused by the absorption of these short wavelengths by the O_3.

Ozone near the ground is a menace, it is a principal constituent of smog. On the other hand, ozone in the stratosphere is a necessity for the maintenance of life. If the short (or ultraviolet) radiation were not absorbed by the O_3, then it would reach the ground, causing mutations in plants and cancer in animals, including humans.

Unfortunately, trace amounts of industrial gases such as chlorofluorocarbons and certain bromine compounds diffuse into the stratosphere from the ground, and by extremely economical reactions, they reverse ozone production, returning O_3 to O_2. The mechanisms are very well understood, and the destruction of O_3 has been measured. Unlike global warming, where the evidence is just beginning to appear, there is no doubt whatsoever that CFCs and other anthropogenic gases have been the principal reason for the destruction of the ozone layer. There is now a worldwide ban on the use of CFCs, yet other molecules that are harmful to the ozone layer are not yet banned. Ozone destruction in the stratosphere is a prime example of extremely small numbers of constituents producing a profound, global effect.

7.3 The Difficulty of Making Predictions

Our discussion of the increase of both greenhouse gases and of local pollutants suggests we should be deeply concerned about the future of the atmosphere, and indeed the environment in general. Figure 7.8 summarizes the situation. From 1860 to 1991, the global rate of energy usage (other than human power) increased from approximately 3.8×10^{18} J/yr to 3.5×10^{20} J/yr. For this ninetyfold increase in energy consumption, the human population increased by less than a factor of 5. So over the past 130 years energy consumption has outpaced population growth by nearly a factor of 20. The rate of energy consumption per capita has gone up from around 80 W in 1860 to nearly 2 kW today.

The value of 2 kW per capita is the total rate of energy used in the world divided by the world population. Each individual uses far less than this because much energy is used to run factories and transportation systems and so on. Yet it is a very good measure of energy consumption. For example, in the United States the average rate of energy consumption per capita is around 10 kW, whereas in poor African countries it is less than 1 kW. Because individual well-being is at present closely linked to per capita energy consumption, it is felt that the poorer countries will have to considerably raise their energy consumption per capita in order to increase the standard of living of their people. We will return to this point in Section 7.4.

Let us look more closely at fossil fuel consumption. Figure 7.9 shows the number of metric tons of carbon emitted into the atmosphere since 1860 due to fossil fuel burning. It includes the burning of solid, liquid, and gaseous hydrocarbons used in all processes (industry, heating, transportation, and so

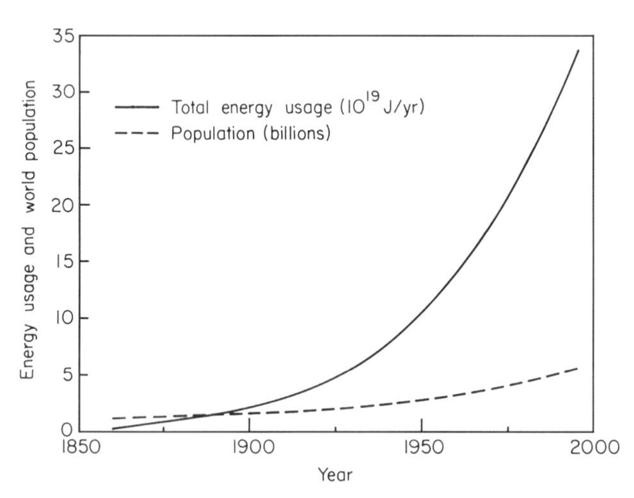

Figure 7.8 The world population and total energy usage since 1860. These graphs have been smoothed. (After Cohen, 1995.)

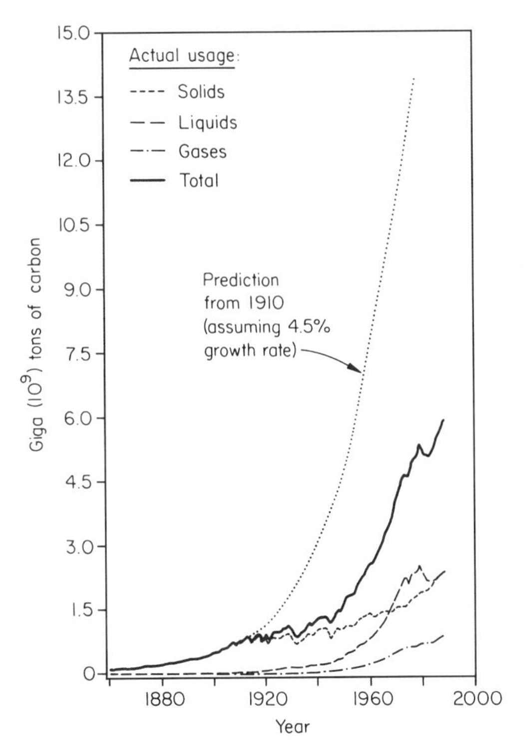

Figure 7.9 The carbon emissions into the atmosphere as a result of fossil fuel burning since 1860. (After "Trends '93," CO_2 Information Analysis Center.)

forth). Its broad trend, of course, follows that of the total world energy use (Figure 7.8). However, Figure 7.9 is more detailed; it has not been smoothed.

Consider yourself to be an engineer in, say, 1910. You would look back at the fossil fuel usage from 1860 and attempt to extrapolate into the future. You would notice that from 1860 to 1910, the graph (Figure 7.9) has a slight upward curvature (it is nonlinear) and so you might think that a compound interest, or exponential model, would be appropriate. The formula for compound interest is

$$F = I(1 + x)^n. \tag{7.9}$$

Here F is the final amount, I is the initial amount, x is the growth per year, and n is the number of years. The amount of fossil fuel burned in 1860 (I) was 0.09×10^9 or 0.09 gigatons (GT) per year. By 1910 it was approximately 0.81 GT/yr (Figure 7.9). By trial and error I have found that a growth rate of 4.5 percent ($x = 0.045$) fits the curve from 1860 to 1910 quite well. In Figure 7.9 I have extrapolated this curve. Evidently the engineer of 1910 would have predicted that by 1970 we would have burned 11.4 GT/yr. In fact we burned less than half of this; around 4 GT/yr. By 1985 the engineer would have overestimated fuel consumption by nearly a factor of 4. Notice that the rate of our energy consumption from around 1950 to 1975 closely follows the predicted trend. But the curve is shifted to the right. So the engineer had the right idea but was assuming business as usual. What in fact happened is that in 1914 a world war began that lasted 4 years. During this period, although the energy use per soldier was high, the energy use per capita for the rest of the industrialized world decreased significantly due to austerity measures. The result was an approximate leveling off in fuel use. Just as the economy was picking up in the 1920s, there was a worldwide depression. This was closely followed by World War II. Not until around 1950 did the growth rate increase in the way the engineer of 1910 had predicted. Notice the bumpiness in the late 1970s and early 1980s. This was due to sudden rises in world oil prices as a result of political decisions by some oil producing countries. It was not due to a worldwide fuel shortage. We will return to this particular incident.

Figure 7.9 shows that prediction is very difficult indeed. War, depression, and artificially imposed fuel shortages can reduce the rate of burning of fossil fuels. Some of you may find comfort in this. You may say that planning for the future is extremely difficult because of our inability to predict social trends. You would be right. But note that in spite of the wars, depressions, and oil price rises, the fact remains that energy usage has outpaced population growth by nearly a factor of 20 since 1860. So, although we should be hesitant about predicting detailed trends, we can still ask a somewhat more modest question: Do we expect that the world energy consumption will increase, decrease, or remain the same during your lifetime as an engineer (the next 40 years or so)?

7.4 A Crude Estimate of Future Energy Consumption

In the previous section we stated that the world energy consumption rate per capita is approximately 2 kW. In other words, the total amount of energy used per year at present is 5.6×10^9 (total world population) $\times\ 2 \times 10^3$ (W) $\times\ 365 \times 24 \times 60 \times 60$ (s/yr) $= 3.5 \times 10^{20}$ J/yr. For such large numbers we use the unit of exajoules (EJ). $1\ \text{EJ} = 10^{18}$ J. Thus, the world is presently using around 350 EJ/yr. Approximately 90 percent of this is due to fossil fuel burning.

While the average per capita energy consumption rate is 2 kW, there is a great difference between the well-developed countries and the less developed countries. For approximately 20 percent of the world population the average per capita energy consumption is 7.5 kW, whereas for the rest of the world, the poorer countries, it is close to 1 kW. It has been estimated that if the poorer countries are to provide adequate medical treatment, schooling, housing, transportation systems, and so on for their people, then their per capita energy usage rate will have to increase to about 3 kW.

Let us assume that in the next 40 years, the less developed countries increase their energy consumption to 2 kW. Let us also assume that the more developed countries decrease their energy consumption to 3 kW per capita. This would mean a per capita consumption decrease by a factor of 2 for the West Europeans and a factor of 3 for the United States. This is unlikely, but let us be optimistic for the moment. Before we can determine the world energy use under these conditions we must estimate the future world population. At present the world population growth rate is 1.6 percent per year. Let us assume a most radical downward trend in growth rate to, say, 0.5 percent per year. This is the approximate growth rate of the well-developed countries. (The poorer the country, the higher the growth rate. For Africa it is presently close to 3%.) So assuming, say, 6.0×10^9 people in 2000, then 40 years later the compound interest formula (Equation (7.9)) yields a world population of 7.3×10^9. If 20 percent of these people (the affluent ones) use 3 kW and the remaining 80 percent use 2 kW, the world energy consumption in 2040 will be 507 EJ. This is nearly 1.5 times the present value. It is based on the assumptions of a dramatic and instantaneous (and therefore unrealistic) drop in world population growth rate and an unrealistic decrease in energy usage in the developed world. It also assumes a modest increase in energy usage for the less developed countries. Without it, large-scale poverty will most likely continue.

You may wish to look at alternative scenarios. In particular, can you determine a situation in which energy usage will feasibly decrease significantly on a global scale? It is difficult to contrive one. (See the article by J.P. Holdren in the general references at the end of this book.)

7.5 The Engineer's Dilemma

The young engineer can expect a significant increase in world energy usage in his or her lifetime. Of course we could revert to a more primitive way of living, or there could be a major cataclysm such as a meteor impact or a large-scale nuclear war. But short of a catastrophe that greatly surpasses the mass carnage of all the wars of the twentieth century, the population and the global rate of use of energy will increase. This will lead to increases in greenhouse gases so that global climate change will most likely occur. It will also lead to increases in local pollution, affecting whole regions.

This presents a great dilemma. For a country to develop, industrialization is needed. Factories and roads must be built. Environmental degradation inevitably follows. We see this in present-day China and Mexico. Yet as the standard of living goes up, the people become more and more aware of the environment and so measures are taken to clean it up. This occurred in the United States in the 1970s and 1980s. Before that, rivers were used as garbage dumps and smoke stacks belched their fumes into populated areas. So it appears that the local environmental quality tends to follow a U-shaped curve: it decreases as industrialization takes place and then increases as the local population becomes wealthier and more conscious of its surroundings. Yet this U-shaped curve mainly pertains to local pollution – pollution that can be seen and directly felt. The global issues, particularly the prospect of global warming, are less immediate, but their effects could be even more profound. And the consequences of global warming will most probably be greater for the poorer, less developed countries than for the richer ones. In the poorer countries there are fewer reserves of food and other resources, so droughts will have an immediate and devastating effect. On the other hand, the richer countries can more easily adapt. So although there is a desire and need to increase the living standards of the less developed countries, the cost of doing so could, in the long run, be greater than the gains.

What are the alternatives? In Chapter 6, I showed that in principle wind and solar energy could provide a significant amount of energy for power plants. Batteries or fuel cells could supply the energy for transportation. And then there is the promise of nuclear energy, be it a cleaner version of fission or the discovery of viable nuclear fusion. If used wisely, we may suppose that these alternatives will alleviate our environmental problems.

Yet even in a hypothetical world free of fossil fuel burning, pollution and environmental degradation will not cease. Batteries require lead, a highly toxic metal, and fuel cells use hydrogen, a much more volatile gas than petroleum. Producing, transporting, and storing it would present severe problems. And nuclear power plants *do* generate toxic waste. Large-scale implementation of nuclear or any other new technology will inevitably produce leakage and accidents, large and small. How devastating these will be depends on a host of factors and cannot be predicted. Even windmills and solar energy plants would bring problems. Their production on a large scale requires new factories, materials, and manufacturing techniques. This will strain resources and particularly affect the economics of developing countries. At least while these technologies were being implemented there would be little improvement in the quality of the environment.

New forms of power generation on a large scale are unlikely in the near future for other reasons. Just as a soldier in battle is unconcerned with the long-term effects of smoking, so, too, a country trying to raise the standard of living is not particularly concerned with a hazy atmosphere, let alone the possible prospects of drought or famine due to global climate change. The cheapest fuel resources will do, and existing technology will be much more attractive than uncertain and more expensive alternatives. In China, which has 20 percent of the world population, the cheapest and the most abundant energy source is coal. Even in the United States, one of the most attractive future commercial power sources is the gas turbine (Section 3.5). It will be powered by natural gas or other hydrocarbon fuels. Although more efficient, it will still add to the global greenhouse gas budget.

There is another issue too. Large corporations prefer a business-as-usual approach, with incremental change. Even small changes require large investment. Typically, an automobile company spends tens or hundreds of millions of dollars developing a new model car and aircraft companies spend billions on new planes. This is for products that depart only in minor ways from their precursors. Usually no radically new principles are involved. You may have noticed the reluctance of automobile manufacturers to produce electric cars. Yet in order to reduce pollution due to hydrocarbons we must change not only the way we power our vehicles but also the way we produce the power to charge them with and the power we use to manufacture them.

How should the engineer respond to this? In former times engineering was synonymous with unconstrained large-scale development. What is technological progress in a world that is constrained by local and global environmental problems? There is no clear answer to these difficult questions. They do, however, suggest that our way of thinking and doing engineering needs to change.

7.6 The Challenge

Incremental engineering dominates the research and development done in most of our automobile and utilities companies. There will always be a need for advances of this kind. For example, without the gains in emission control achieved over the past 30 years (Figure 7.2), there is no question that the atmospheric loading of NO_x and other constituents would have increased dramatically. At least in the United States they have been held approximately steady (Figure 7.3). Yet as we have shown in Sections 7.2–7.4, if the world population increases at even a modest rate, and the less developed countries attempt to increase the standard of living of their people, the incremental gains will be swamped by net increased energy usage.

These considerations may suggest that the engineer should become radical, shunning work on incremental advances and engaging in developing new energy sources and transportation systems. But as I have suggested, radical changes are difficult to implement and even clean energy sources will bring new problems. Hydroelectricity is a case in point. Small plants are clean and blend into the environment. However, at a large scale they can produce social problems and undesirable ecological and environmental change. We must explore new technologies, as well as incrementally advance traditional ones. Most important, engineers must begin to ask questions about the overall purpose of their work. In the old days when progress meant bigger, faster, and more powerful, such questions were less necessary.

Asking questions implies dialogue, not only with one's fellow engineers but with other members of society. In the past many of the innovations in engineering were fostered by the military. Some of the these advances have affected society as a whole, but most have been employed in devices that are useful only for war. In developing these devices, there has been very little dialogue between the engineer and the public; the defense agencies have determined the problems, and the engineers have willingly solved them. In an overcrowded world threatened by environmental destruction, not only do the engineering problems change, but the very nature of the way they are defined must change also. The engineer should think less of him- or herself as a servant of corporations and government and more as a technical expert who can aid in policy-making, and mold public opinion. Indeed, the engineer's privileged education implies an ethical obligation to do so. As experts they can no longer afford to keep to themselves, at the edge of the campus.

The problems are immense, yet there is room for optimism. In my discussion of the world carbon emissions (Section 7.3), I drew your attention to their approximate leveling off in in the late 1970s (Figure 7.9). This was due to reduced fossil fuel use as a result of the oil embargo after the Arab–Israeli war of 1973. The United States, which uses around 20 percent of the world's energy, played a major role here, significantly increasing energy efficiency in transportation, industry, and housing. (Alas, with the reduced oil prices in the 1980s much of this good work has become undone (Figure 7.9).) This shows that concerted action *can* have an effect on the global environmental problem. The reduction in CFC emissions provides another example. But it is irresponsible to respond to problems only when the situation becomes intolerable. Indeed many specialists feel that by the time global climate change becomes clearly apparent we will not be able to avert massive suffering.

There are no neat solutions to the problems I have raised in this book. It is unlikely we will find a new energy source that will change everything. Clearly, we must study new concepts and designs and keep trying to achieve incremental advances in pollution control and engine efficiency. Above all we must discuss the issues we have raised here with the population at large and become much more concerned with their education. There has been a tendency in the past for engineers and scientists to mystify their profession. But the problems dealt with here are societal, and until the general public sees the technological issues clearly and becomes involved, progress will be slow. The consequences will affect us all. The analysis I have outlined suggests that the global energy use must be reduced. This means the notion of technological progress must be carefully reevaluated. This will inevitably lead to a reevaluation of our way of life.

Appendix: The Equations of Fluid Motion

Figure 3.21 of Chapter 3 shows the velocity vectors for the motion inside an engine. There I stated that these were the solution of the equations of fluid motion. Again in Chapters 4 and 5 I alluded to these equations. I did not derive them because they are quite complicated. However our study of heat transfer and fluid mechanics has provided us with enough background to develop them in an intuitive way, by analogy with the heat conduction equation.

We recall that the heat conduction equation tells us that the rate of change of temperature with time is proportional to the second spatial derivative of temperature. For one dimension this is (Equation (4.10))

$$\frac{\partial T}{\partial t} = \alpha \frac{\partial^2 T}{\partial x^2}.$$

This equation was derived using Fourier's law, which states that the heat transfer is proportional to the temperature gradient. Now, in Chapter 3 we found that the momentum transfer is proportional to the velocity gradient (Equation (3.6)). Thus, by analogy to the heat conduction equation we may be tempted to write

$$\frac{\partial \mathbf{V}}{\partial t} = \nu \frac{\partial^2 \mathbf{V}}{\partial x^2}. \tag{A.1}$$

This states that the momentum per unit mass of a fluid will change in a way that is proportional to the comparative surplus or deficit of momentum at some position. If there is a region of high momentum, its tendency will be to diffuse to regions of lower momentum. However, unlike temperature, momentum transfer can also be affected by a pressure difference, or gradient. This suggests that Equation (A.1) should be modified to take the form

$$\frac{\partial \mathbf{V}}{\partial t} = \nu \frac{\partial^2 \mathbf{V}}{\partial x^2} - \frac{1}{\rho}\frac{\partial p}{\partial x}. \tag{A.2}$$

Notice that this Equation (as well as Equation (A.1)) is a force balance: all of the terms have the units of acceleration, or force per unit mass. If we assume the fluid is frictionless ($\nu = 0$), Equation (A.2) becomes

$$\frac{\partial \mathbf{V}}{\partial t} = -\frac{1}{\rho}\frac{\partial p}{\partial x},$$

that is, if there is a pressure gradient in a fluid, there will result an acceleration providing there are no restraining forces present. (Note the minus sign because the acceleration will be from high to low pressure.) However, in a real fluid (Equation (A.2)) the acceleration will be modified by the presence of viscous forces.

Now, if we focus on a point in a fluid, a so-called fluid particle, we will notice that its velocity is not only a function of time but also a function of position, that is, $\mathbf{V} = \mathbf{V}(x, t)$. Using the concept of partial differentiation, we may write

$$d\mathbf{V} = \frac{\partial \mathbf{V}}{\partial x}dx + \frac{\partial \mathbf{V}}{\partial t}dt, \tag{A.3}$$

that is, a small change in velocity is due to the sum of the contribution of changes in velocity in space as well as time. (The flow going through the contraction (Section 3.3.1) is a good example of the spatial change of velocity.) Dividing Equation (A.3) by dt we find

$$\begin{aligned} \frac{d\mathbf{V}}{dt} \equiv \mathbf{a} &= \frac{\partial \mathbf{V}}{\partial x}\frac{dx}{dt} + \frac{\partial \mathbf{V}}{\partial t} \\ &= V_x\frac{\partial \mathbf{V}}{\partial x} + \frac{\partial \mathbf{V}}{\partial t}, \end{aligned} \tag{A.4}$$

where $V_x \equiv dx/dt$ is the velocity in the x direction. This is the equation for the acceleration at any position within a fluid. Clearly, Equation (A.2) must be further modified. Its left-hand side is the acceleration of a fluid particle, but we have just shown that this consists of two terms, not the single term we had assumed. Equation (A.2) thus becomes

$$\frac{\partial \mathbf{V}}{\partial t} + V_x\frac{\partial \mathbf{V}}{\partial x} = -\frac{1}{\rho}\frac{\partial p}{\partial x} + \nu\frac{\partial^2 \mathbf{V}}{\partial x^2}. \tag{A.5}$$

Finally, we must extend this to three dimensions:

$$\begin{aligned} \frac{\partial \mathbf{V}}{\partial t} + V_x\frac{\partial \mathbf{V}}{\partial x} + V_y\frac{\partial \mathbf{V}}{\partial y} + V_z\frac{\partial \mathbf{V}}{\partial z} &= \nu\left[\frac{\partial^2 \mathbf{V}}{\partial x^2} + \frac{\partial^2 \mathbf{V}}{\partial y^2} + \frac{\partial^2 \mathbf{V}}{\partial z^2}\right] \\ &\quad -\frac{1}{\rho}\left(\mathbf{i}\frac{\partial p}{\partial x} + \mathbf{j}\frac{\partial p}{\partial y} + \mathbf{k}\frac{\partial p}{\partial z}\right) - g\mathbf{k}. \end{aligned} \tag{A.6}$$

Here V_y and V_z are the velocities in the y and z directions in the rectangular coordinate system (with z upward), and **i**, **j**, and **k** are unit vectors in the x, y, and z directions respectively. Because Equation (A.6) is a force balance, we have added the gravitational acceleration for the vertical direction. Note that if $\mathbf{V} = 0$ (no motion), Equation (A.6) reduces to the hydrostatic condition (Equation (5.2)).

Equation (A.6) is a vector equation because **V** is a vector. It is known as the Navier–Stokes equation. It is a force balance equation, stating that the acceleration per unit mass of the fluid at any position in the flow is equal to the sum of viscous, pressure, and gravitational forces acting at that position. Because each point in a fluid is in contact with its neighbors, a change of pressure or velocity at one position affects the surrounding velocities and pressures, and those affect their neighbors, and so on. For this reason the Navier–Stokes equation is called a field equation, and in this respect it is like the equations describing electric and magnetic fields. Equation (A.6) can easily be extended to include other forces such as forces due to the rotation of the whole fluid body (Coriolis force). It is very general and applies to most engineering fluids such as oil, air, and water and can be easily modified for fluids such as mercury or polymers. It holds for turbulent as well as laminar flow: the left-hand side takes into account all of the jerky, erratic motion of a turbulent fluid. For example, if we focus at one position of a waterfall and measure at one instance every term on the left-hand side (this means measuring not only the velocity vector but its rate of change with time and space), and at the same time measure the pressure gradient, and all of the other terms on the right-hand side, they will balance. We know this because it has been experimentally checked for waterfalls, jet engine exhausts, and hurricanes, as well as the flow in bearings, pipes, and chimney stacks. Although we know it is correct because it always balances, we do not know how to solve it, except for some very simple cases. We cannot solve it to tell us how a turbulent eddy will evolve. Solving the equation means determining **V** as a function of position and time everywhere in the flow.

The fact that we can derive an equation but not solve it must seem a strange state of affairs. It is as if we had written a book that we cannot read! A large part of the difficulty is due to the nature of the terms on the left-hand side: the velocities multiplied by their gradients. These terms are responsible for the much greater stresses that occur in turbulence and for the way a fluid undergoes transition. They are also responsible for the chaotic nature of fluid flow at high Reynolds numbers.

Although we cannot solve the Navier–Stokes equations in a nice analytical way, we can use a computer to do a numerical approximation, or a direct numerical simulation, as it is called. The equations have five dependent variables; three components of velocity, pressure, and density. (An extra two equations for state and mass conservation are used to close the set; note that the Navier–Stokes equation is a vector equation for the three components of velocity.) At some time $t = t_0$ these variables are specified for every one of the thousands or more positions of relevance in a particular flow, for example, at all the grid points of the engine example of Figure 3.21. Because their positions are close together, the velocity and pressure gradients can also be determined. Because the Navier–Stokes equations must balance, this information can then be used to determine the rate of change of the velocity components, $\partial \mathbf{V}/\partial t$. The rate of change is determined over a finite time step, Δt, as $\Delta \mathbf{V}/\Delta t$. Because we know the velocity at $t = t_0$, it can now be determined at $t = t_0 + \Delta t$ as $\mathbf{V} = \mathbf{V}_0 + \Delta \mathbf{V}$. Thus, the velocity field is predicted over the small period Δt. The program is then time-stepped another Δt, and all of the components are reevaluated, and so on until there are a sufficient number of steps to observe the required evolution of the flow field. In the case of an engine Δt may need to be as short as 1 ms in order to capture the rapid changes. The computer program may evolve for 1 s or 1,000 time steps. For making numerical predictions in the atmosphere, Δt may be minutes or even hours; the latter if only large-scale events are of relevance. Such calculations require the very largest computers, and the computation time is great; it may take hours or even days to compute what is occurring in seconds inside an engine. However, it is the best we can do at the moment. Maybe some day a general way of solving the Navier–Stokes equations will be found or a new approach to fluid mechanics will be discovered. Or maybe computers will become so large that it will not worry us that this approach lacks elegance. The subject is important and the prospects are exciting.

Finally, it must be emphasized that my development of the Navier–Stokes equation has been unrigorous, although Equation (A.6) itself is correct. It has been done only to let you have a peek, as it were, of what is in store for later on. You will derive them in a more formal way in the third or fourth year, and there you will also solve them for laminar flow in a pipe and show that the solution is indeed a parabola as was stated in Chapter 3. Some of you will study them in much greater detail in graduate school, where you will also find that a number of your professors have spent their career probing their mysteries.

General References

The following provide further background at the undergraduate level:

Chapter 2

Moran, M.J., and Shipiro, H.N. (1992). *Fundamentals of Engineering Thermodynamics*, Wiley.
Wark, K. Jr. (1988). *Thermodynamics*, McGraw Hill.

Older texts that are particularly clear and concise are
Feynman, R.P., Leighton, R.B., and Sands, M. (1963). *The Feynman Lectures on Physics*, Addison-Wesley. (This has wonderful chapters on Thermodynamics, Fluid Dynamics, and many other topics.)
Keenan, J.H. (1941). *Thermodynamics*, Wiley.
Reynolds, W.C., and Perkins, H.C. (1970). *Engineering Thermodynamics*, McGraw Hill.
Spalding, D.B., and Cole, E.H. (1958). *Engineering Thermodynamics*, Arnold.

Chapter 3

Sabersky, R.H., Acosta, A.J., and Hauptmann, E.G. (1989). *Introduction to Fluid Mechanics*, Macmillan.
White, F.M. (1986). *Fluid Mechanics*, McGraw Hill.

A slightly more advanced book with an environmental emphasis is
Fay, J.A. (1994). *Introduction to Fluid Mechanics*, MIT Press.

Chapter 4

Incropera, F.P., and DeWitt, D.P. (1985). *Fundamentals of Heat and Mass Transfer*, Wiley.

Chapter 5

Graedel, T.E., and Crutzen, P.J. (1993). *Atmospheric Change: An Earth System Perspective*, W.H. Freeman and Co.
Hess, S.L. (1959). *Introduction to Theoretical Meteorology*, Holt, Rinehardt and Winston.

Chapter 6

Kraushaar, J.J., and Ristinen, R.A. (1993). *Energy and Problems of a Technical Society*, Wiley.
Kreith, F., and Kreider, J.F. (1978). *Principles of Solar Engineering*, Wiley.
Fay, J.A. (loc. cit.)

Chapter 7

Section 7.2

Greene, D.L., and Santini, D.J. (eds.) (1993). *Transportation and Global Climate Change*, American Council for an Energy-Efficient Economy.
Graedel et al. (loc. cit.)
United Nations Environmental program. (1987). *The Greenhouse Gases*, UNEP/GEMS Environmental Library No 1.

Section 7.3

Cohen, J.E. (1995). *How Many People Can the Earth Support*? W.H. Norton and Co.

Section 7.4

Scientific American (September 1990). *Energy for Planet Earth.* (The article by J.P. Holdren (p. 157) discusses various population and energy use scenarios.)

Specialized Sources

In addition to the general references, I acknowledge the following more specialized sources:

Arrow, K., Bolin, B., Costanza, R., Dasgupta, P., Folke, C., Holling, C.S., Jansson, B-O., Levin, S., Mäler, K-G., Perrings, C., and Pimental, D. (1995). Economic Growth, Carrying Capacity and the Environment, *Science*. **268**, 520–521.

Calvert, J.G., Heywood, J.B., Sawyer, R.F., and Seinfeld, J.H. (1993). Achieving Acceptable Air Quality: Some Reflections on Controlling Vehicle Emissions, *Science*, **261**, 37–45.

Cohen, J.E. (1995). Population Growth and the Earth's Human Carrying Capacity, *Science*, **269**, 341–346.

Gipe, P. (1995). *Wind Energy Comes of Age*, Wiley.

Houghton, J.T. (1986). *The Physics of Atmospheres*, Cambridge University Press.

Johansson, T.B., Kelly, H., Reddy, A.K.N., Williams, R.H., and Burnham, L. (eds.) (1993). *Renewable Energy*, Island Press.

Khalighi, B., Haworth, D.C., and Huebler, M.S. (1994). Multidimensional Port-and-in-Cylinder Flow Calculation and Flow Visualization Study in an Internal Combustion Engine with Different Intake Configurations. *SAE Technical Paper*, 941871. (Presented at Fuels & Lubricants Meeting & Exposition. Baltimore, Maryland. October 17–20, 1994.)

Kravchik, T., Sher, E., and Heywood, J.B. (1996). From Spark Ignition to Flame Initiation, *Combustion Science and Technology*, **108**, 1–31.

Krenz, J.H. (1976). *Energy Conversion and Utilization*, Allyn and Bacon, Inc.

Mitchell, J.F.B. (1989). The "Greenhouse" Effect and Climate Change. *Rev. Geophysics*, **27** (1), 115–139.

Monin, A.S. (1972). *Weather Forecasting as a Problem of Physics*, MIT Press.

Ramanathan, V., Barkstrom, B.R., and Harrison, E.F. (May 1989). Climate and the Earth's Radiation Budget. *Physics Today*, American Institute of Physics, 22–32.

Ramos, J.I. (1989). *Internal Combustion Engine Modelling*, Hemisphere.

Scorer, R.S. (1978). *Environmental Aerodynamics*, Ellis Horwood.

Seinfeld, J.H. (1989). Urban Air Pollution: State of the Science, *Science*, **243**, 745–752.

Shepherd, D.G. (1990). *Historic Development of the Windmill*, NASA Contractor Report 4337.

Spera, D.A. (ed.) (1994). *Wind Turbine Technology*, ASME Press.

Sutton, O.G. (1953). *Micrometeorology*, McGraw Hill.

Tennekes, H., and Lumley, J. L. (1972). *A First Course in Turbulence*, MIT Press.

Trends '93. (1994). CO_2 Information Analysis Center, Oak Ridge National Laboratory.

Turns, S.R. (1996). *An Introduction to Combustion*, McGraw Hill.

Wayne, R.P. (1991). *Chemistry of Atmospheres*, Clarendon Press, Oxford.

Answers to Problems

Chapter 1

1.1 0.186 ppmv (approximately half the vehicle emissions for that year).
1.2 0.021 ppmv.
1.3 (a) 2, 1. (b) 1.27 kg CO, 1 kg CO_2. (c) 3 kg CO_2.
1.4 11.7 kW, 0.65 ppmv/yr.
1.5 car, 365 kJ; olive oil, 566 kJ (same order of magnitude).
1.6 A good guess is 10 m/s or approximately 20 mph (see Chapter 5). Therefore, the total kinetic energy is 2.6×10^{20} J. This is equivalent to 1,200 of the largest nuclear bombs ever made.

Chapter 2

2.1 Both are equally pure (or impure).
2.2 (a) $\Delta U = 6{,}500$ J. (b) $\Delta E = -6$, $\Delta E = +6$, $W = +3$.
2.3 (a) wood ball $Q=0$, W – ve. Therefore, ΔE – ve. (Instead of causing the surrounding temperature to rise, system could have lifted a weight external to it and still undergone the same process. This is consistent with the formal definition of work.) (b) copper ball $Q=0$, $W=0$ (no effect on surroundings), hence, $\Delta E = 0\,(PE \rightarrow$ internal energy).
2.4 (a) $W=-9.8$ J, $Q=0$, $\Delta E=-9.8$ J. (b) (i) insulated, $W=0$, $Q=0$, $\Delta PE =$
$-\Delta U$. Thus $\Delta U = +9.8$ J. (ii) $Q = \Delta U = -9.8$ J.
2.5 Ratio, $2ln2 = 1.39$, ΔE is the same for both processes. $W-$ve, $\Delta U = 0(\Delta T = 0$ and perfect gas); hence, Q is positive ($W + Q = 0$). Also, $Q_{(a)} > Q_{(b)}$ since $|W_{(a)}| > |W_{(b)}|$.

2.6 86 m.
2.7 $p_2 = 0.5 \times 10^5$, $T_2 = T_1$
2.9 $\eta_{\text{Otto}} = 1 - 1/(r^{\gamma-1})$, $\eta_{\text{Diesel}} = 1 - [1/(r^{\gamma-1})][(r_c^{\gamma} - 1)/\gamma(r_c - 1)]$, where $\gamma = c_p/c_v (> 1)$, $r = p_2/p_1$, and $r_c = v_2/v_1$. $\eta_{\text{Diesel}} < \eta_{\text{Otto}}$.
2.12 Relative temperature difference is smaller for Venus.
2.14 (a) no. (b) yes. (c) A.

Chapter 3

3.1 My estimates are 10^5, 10^8, 10^9, 10^9, 2.1×10^6, and 10. Yours may differ by an order of magnitude or so.
3.2 For the table, the depth is 5 mm when flow becomes turbulent at $x = 0.33$ m. For the aircraft, the depth is 0.08 mm at $x = 5.33$ mm.
3.3 25 m/s.
3.4 233 N; 467 W; $Re = 6$ (laminar).
3.6 2.23 m/s; 1.12×10^6 (turbulent); 49.2 Pa/m; 4.92×10^7 Pa for full length; 0.43 MW.
3.7 $\dot{m}_{\text{in}} = 4.8 \times 10^{-3}$ kg/s; $\dot{m}_{\text{out}} = 3.88 \times 10^{-3}$ kg/s; therefore, a leak.
3.8 299.2 K. The kinetic energy increases.
3.9 0.577 kg/s.
3.10 $\dot{m} = 2.03$ kg/s; $V_{\text{in}} = 0.196$ m/s; $V_{\text{out}} = 2.15$ m/s.
3.11 (a) $T_2 = 489$ K; $T_3 = 1{,}000$ K, $p_3 = p_2 = 6 \times 10^5$; $T_4 = 599$ K, $p_4 = p_1$. (b) Compressor work $= 1.96 \times 10^5$ J/kg. (c) turbine work $= 4.01 \times 10^5$ J/kg.
(d) $\eta = 40\%$.
3.12 10^{11} J; 25 min; 72 MW.
3.13 $\tau = 2\pi\sqrt{\ell/g}$.
3.14 For volcano: $\eta = 7.82 \times 10^{-4}$ m, $Re_\ell = 6 \times 10^8$. For cloud: $\eta = 5.6 \times 10^{-4}$ m, $Re_\ell = 10^7$. $\ell/\eta \propto (Re_\ell)^{3/4}$. You can show this analytically using the definition of ϵ and η.

Chapter 4

4.1 125 kW_{in}; 165 kW_{out}.
4.2 $T - T_H = (\dot{Q}_0/Aka)[e^{-ax} - 1]$, where A is the area of the wall.
4.3 $T - T_0 = -(PL^2/2k)[x^2/L^2 - x/L]$.
4.5 $Ra \sim 10^{11}$ (high, therefore convective); $w \sim 1$ m/s; $\tau \sim 3$ s. Molecular diffusion time ~ 10.4 days, i.e., 3×10^5 longer than turbulent diffusion time.
4.6 3×10^5.

4.7 $Re_L \sim 10^8$; $Nu \sim 10^5$; $H = 31.27$ W/(m^2 K); total cooling rate is 1.9 MW (large).

4.9 Ratio of buoyancy to inertia forces is $g\beta\Delta T\ell/V^2$. For problem 4.7, this ratio is 9.4×10^{-3}. It is small compared to unity, so cooling is by forced convection.

4.10 400π W/m^2.

4.11 5.95×10^6.

4.12 0.42.

4.13 (a) 0.56 kW/m. (b) 1.0 kW/m.

Chapter 5

5.1 (b) 6.9×10^{-3} W/m^2. (c) About an order of magnitude greater than a full moon.

5.2 (a) 2.2×10^{-3} m^2/s^3. For the ABL, the total energy dissipated is ϵ multiplied by the mass of the ABL. It is 1.34×10^{15} W. Thus for the whole atmosphere the power dissipated is $5/3 \times 1.34 \times 10^{15}$ W $= 2.24 \times 10^{15}$ W. This is consistent: The power of the atmospheric motion must be dissipated because it is always being replenished from the sun. (b) $\dot{Q} = mc_p(\Delta T/t)$. If $\dot{Q} = 2.24 \times 10^{15}$ W, $c_p = 1{,}000$ J/kg K and $m = 5.2 \times 10^{18}$, then $\Delta T = 3.7 \times 10^{-2}$ K in 24 hr. This (small) quantity is taken care of in the radiation budget.

5.3 $w \sim \ell(dV/dZ)$; $w = 1.0$ m/s. Note: $w/\ell \sim dV/dZ$ implies that the frequency of the turbulence (both sides have dimensions of t^{-1}) is the same as the frequency of the shear. Thus, they are tuned to each other. This is generally the case if there is only a single forcing mechanism for the turbulence.

5.4 4.7×10^6 J. Fan provides 2.9×10^6 J in 8 hr. It facilitates the forced convection of water vapor from the surface of the water. It does not cause the evaporation.

5.5 (c) $(R_u/M_d)[(1 + q_v/(M_v/M_d))/(1 + q_v)]$. (d) $T^* = [(1 + q_v/(M_v/M_d))/(1 + q_v)]T_0$; 7 K. The virtual temperature, T_*, is always higher than T because water vapor is less dense than air ($M_v/M_d = 0.622$).

5.6 $p = p_0(T/T_0)^{g/R\gamma}$. Thus, $p/p_0 = 0.79$. For the isothermal atmosphere, $p/p_o = e^{-z/H}$, where $H = RT/g$. Here too $p/p_0 = 0.79$, showing that for this height difference the two are the same. (They will depart from each other for larger height differences.)

5.7 (a) 5.2×10^{21} J/K (atmosphere); 4.2×10^{24} J/K (oceans). Thus, the heat capacity of the oceans is approximately one thousand times greater than that of the atmosphere. (b) 1.02×10^7 (J/K m^2). (c) $\tau \sim c_p\rho H/\sigma T^3 \sim 87$

days. (This is to an order of magnitude. In fact the correct answer is closer to 10 days.)

5.8 72.4 days.

5.9 48 cm. Ocean expansion poses the greatest threat to island nations and coastal cities.

5.10 (a) $f = [(g/T_0)(dT/dz)]^{1/2}$. For $dT/dz = 0.02$ K/m, $f = 2.6 \times 10^{-2}\,\mathrm{s}^{-1}$. This corresponds to a period of 38 s. (b) $-\infty$. (c) -1.28×10^{-2} K/m. Only a slightly greater lapse rate than adiabatic for this very mild shear.

5.11 (a) $Ri_B = 9.7 \times 10^{-3}$. Since this is small compared with unity it suggests the effect of stable stratification is negligible here. (b) $Nu = 10^5$, $Ri_B = -6.3 \times 10^{-3}$. Heat flux is upward. To keep ice at $-1°$C, 1.25×10^6 W is needed, but it should not matter if the ice cools.

Chapter 6

6.1 (a) $C_{12}H_{22}$, 11 moles. (b) 0.046, 0.07, 0.66.

6.2 (a) approximately 3 photons/molecule. (b) 1.25×10^8 kg; 13,210 people; 5.11×10^5 km^2 or around 5.5 percent of the United States.

6.3 1.8×10^{11} kg, 0.31 years; for methane: 5.74×10^{12} kg, 3.6 years.

6.4 4.39×10^{21}, $3.22 \times 10^{12}\,\mathrm{s}^{-1}$, 199 years.

6.6 First day, 3.97×10^9 J, second day, 8.03×10^9 J. Implies gustiness is better, but intermittent load is hard to control and can cause structural fatigue.

6.7 (a) 1 km^2, 3,330 people. (b) 63.6 percent. The temperature of the salt is greater than that of the water entering the turbine. Similarly, the lowest temperature of the water in the cycle is still much above ambient. (c) 41.7 kg/s, 100 m/s.

6.8 12.6 W/m^2, 89×89 m^2 compared to 17×17 m^2 on earth.

6.9 25.5 MW.

6.10 (a) 8.3 percent. (b) $\eta_{th} = 2.78$ percent, $\dot{m} = 15.15 \times 10^3$ kg/s, 3.6×10^9 W (heat input), area of heat exchanger 9×10^5 m^2 or 950×950 m (this is very large due to the small temperature difference).

*Index**

*Page numbers in bold type indicate definitions or main entries.

Conversion of Units

Energy (kg m^2/s^2; Joule, J)

1 British thermal unit (Btu)	$= 1.055 \times 10^3$ J
1 electron volt (eV)	$= 1.6022 \times 10^{-19}$ J
1 exajoule (EJ)	$= 10^{18}$ J
1 kilocalorie (= 1 Calorie used in food labels)	$= 4.187 \times 10^3$ J
1 kilowatt hour (kW hr)	$= 3.6 \times 10^6$ J
1 quad	$= 1.055 \times 10^{18}$ J

Energy equivalents (Joule)

1 barrel of crude oil	$= 6.12 \times 10^9$ J
1 barrel of automobile gasoline	$= 5.54 \times 10^9$ J
1 tonne of coal	$= 2.5 \times 10^{10}$ J
1 tonne of TNT	$= 4.184 \times 10^9$ J

Force (kg m/s^2; Newton, N)

1 dyne	$= 10^{-5}$ N
1 pound-force (lbf)	= 4.4482 N

Length (meter, m)

1 foot (ft)	= 0.3048 m
1 mile (mi)	$= 1.609 \times 10^3$ m

Mass (kilogram, kg)

1 pound (lb)	= 0.4536 kg
1 ton (2240 lb)	= 1,016.0 kg
1 tonne (metric ton)	= 1,000 kg

Power (kg m^2/s^3; Watt, W)

1 horsepower	= 745.7 W

Pressure (kg/(m s^2); Pascal, Pa)

1 atmosphere	$= 1.013 \times 10^5$ Pa
1 bar	$= 10^5$ Pa
1 lbf/in^2 (psi)	$= 6.895 \times 10^3$ Pa

Temperature (Kelvin, K)

degree Celsius (C)	$K = C + 273.15$
degree Fahrenheit (F)	$K = (F + 459.67)/1.8$
degree Rankine (R)	$K = R/1.8$